Psychoanalysis and Governance

Psychoanalysis and Governance makes a cogent argument for the use of psychoanalytic perspectives in the understanding of governance, the process of collective decision-making that maintains and reshapes communities.

This book is highly relevant to those interested in the ever-expanding field of applications of psychoanalysis and for all those willing to observe the discursive and affective underpinnings of public policy, administration, and planning. It locates the potential for self-analysis and self-transformation within governance, yet also indicates governance as the confluence of diverging understandings of the ideas of community and governance itself, as the place where competing desires and variegated patterns of fears and hopes collide and hold the transformational potential to destabilize the community.

Building on Freudian, Lacanian, and other psychoanalytic traditions, the book enriches our understanding of governance, the way communities remember and forget, are haunted by the past, remain untransparent to themselves yet also retain the possibility of reinvention, of imagining alternative selves, new futures, and discover paths to move in that direction. This book will be a suitable for psychoanalysts, planners, and all those interested in informed governance.

Kristof Van Assche, a professor at the University of Alberta, is interested in evolution and innovation in governance. He explores the relations between knowledge, affect, and organization in their implications for sustainability thinking and collective strategy. He is one of the proponents of Evolutionary Governance Theory (EGT).

Monica Gruezmacher is an associate researcher at the University of Alberta and an adjunct professor at Memorial University. In the past she worked as a policy adviser. She is interested in the challenges of managing long-term relationships between communities and their environment, especially in rural, remote, and challenging settings.

Psychoanalysis and Governance

Discourse and Decisions, Identities and Futures

**Kristof Van Assche and
Monica Gruezmacher**

Routledge
Taylor & Francis Group

LONDON AND NEW YORK

Designed cover image: © Getty Images

First published 2025
by Routledge
4 Park Square, Milton Park, Abingdon, Oxon OX14 4RN

and by Routledge
605 Third Avenue, New York, NY 10158

Routledge is an imprint of the Taylor & Francis Group, an informa business

© 2025 Kristof Van Assche and Monica Gruezmacher

British Library Cataloguing-in-Publication Data
A catalogue record for this book is available from the British Library

ISBN: 9781032696645 (hbk)
ISBN: 9781032696683 (pbk)
ISBN: 9781032696676 (ebk)

DOI: 10.4324/9781032696676

Typeset in Optima
by Newgen Publishing UK

Contents

Prologue

Why this book? First of all, we believe that psychoanalysis is very much alive. Second, psychoanalysis has much to say about governance, that is, about politics in the very broadest sense, in the sense of everything that goes into the making of collectively binding decisions for a community. If there is binding and community, there is the potential for shaping, even if the community is unaware of it, or rejects the option outright. A third reason is that the literatures on governance, rich and varied they may be, and crossing the disciplines of politics, sociology, public administration, organization studies, and planning, tend to veer back to modernism, to a set of ideas originating in the eighteenth century, and envisioning governance as finding rational solutions for objectively existing problems. Dreams and desires, fear and hope, unconscious motivations, and the power of stories and governance to create and reject realities are immediately out of the picture.

The book is not an encyclopedia of governance, nor an application of one style of psychoanalysis to core issues of governance. Rather, we present an enriched perspective on governance, where psychoanalysis is at play in different ways. Concepts such as the unconscious are introduced into governance theory, while more traditional concepts such as actors, learning, and informality are refracted through a psychoanalytic lens. That lens might rather be compared to a stained-glass window, as influences of more than one tradition can be discerned. Sigmund Freud and Jacques Lacan are the main ones, however, and their interpreters in politics, planning, administration, philosophy, anthropology, and geography. We are much indebted to the work of Jason Glynos and Yannis Stavrakakis, Jean Hillier and the late Michael Gunder, Stephen Frosh, and Derek Hook, as well as Stijn Vanheule, Paul Moyaert, and of course, the mercurial presence of Slavoj Žižek.

Readers will notice some curious characters woven into the debate: organization theorists such as David Seidl, Mats Alvesson, Martin Kornberger, and the late Barbara Czarniawska. They are essential for our argument, as their stories about the level between individual and community, and about the processes of organizing outside organizations, help explain how thinking and organizing co-evolve, and how multilevel structures can emerge and function. Through them, and via concepts borrowed from evolutionary governance theory, a distinct influence of systems theory can be traced, with Niklas Luhmann as the central figure. Luhmann's theory of ever differentiating social systems, functioning based on procedures enabling focused observation but also introducing new opacities, provides unifying concepts such as system, environment, and co-evolution, and, even if he did not devote much attention to people, whom he labeled "psychic systems," for us his centrality of meaning as a medium for both individuals and social systems to communicate and to form themselves, rang very true, very much akin to psychoanalysis, and fundamentally helpful in exploring the use of psychoanalytic ideas for the understanding of governance. More precisely, a Luhmannian-inspired governance theory, can, we argue, accommodate insights from psychoanalysis more easily than others, and can, we believe, be the basis for the formation of psychoanalytic perspectives on governance.

The weft connecting governance concepts anew, is then a narrative one, with meaning created, affect recognized and produced by narrative, with unconscious desire, narratives, and Master Signifiers becoming thinkable, and with interpretation and reinterpretation becoming the bread and butter of organizing, providing the essential linkage between thinking and organizing. Both psychoanalysis and systems theory, thus organization and governance theories inspired by it, recognize the gradual creation of identity in processes of co-evolution, where individuals, communities, and their organizations slowly become possible, acquire a character, a delineation over time, and in the co-presence of others, other people, and their stories.

The resulting perspective on governance, however, is psychoanalytic more than anything else. Luhmann, together with the more psychoanalytically inclined interpreters of governance, provides powerful assistance in bringing governance theory to a regrounding in a Lacanian fashion, emphasizing negativity at the heart of governance, a gap between governance and community that makes both possible. Communities can develop governance systems that enable them not only to represent themselves to themselves but that allow them also to envision alternative futures and transform themselves in that imagined direction. The community cannot recognize itself entirely in governance, as governance can never entirely do what it promised. A web of tensions and interpretative differences is bound to appear as governance cannot coincide entirely with community, and the difference introduces all phenomena associated with language, desire, and identification.

We do not intend this book to be the last word; instead, we wish it to provide food for discussion, or better, we hope it is an invitation to further explore the terrain we have just started to map out. We believe it represents a step forward in the construction of governance theories giving a fuller account of their parts and relations, toward a redescription of the potentialities of governance. Starting from a constitutive lack might help resist the lure of modernism, which remains alluring for academics, bureaucrats, and policymakers alike, and which, alas, does not bring back the repressed, but revives fantasies of optimal governance structures, processes, and results that then become codified in normative theories of governance not that conducive to reflexivity.

We also hope that readers can enjoy themselves going through the pages of this book. We wish to add a modest warning: we aimed for an engaging tone and accessible treatment of what is inherently complex, so it might prove rewarding to stick around for a while. The story unfolds slowly and steadily, chapters build on each other, and reading the last chapter, we believe, is a different experience after following the path.

The authors

1   Introduction

Governance and psychoanalysis

Governance

Nothing new to see here. Governance has always existed. As long as people have lived together, decisions have had to be taken. Sometimes it was about addressing the situation at hand. Other decisions had the character of rulemaking, the crafting of rules to guide the community in later situations that were similar enough to fall under the same rule. In this book, we speak of governance when collectively binding decisions are taken. For governance to stabilize itself, it must produce rules, but also guide itself through rules. Those guiding rules are often not produced in the community itself. More commonly, they were taken a long time ago, possibly at the founding of the community and elsewhere, perhaps in a capital city (Anderson, 2006; Scott, 2017). Elsewhere in this case tends to be a center of power, of higher-level governance. This leads us to introduce the idea of multilevel governance, which is indeed as old as societies themselves. Except for the most ancient or remote communities, governance tends to organize itself by distinguishing spatial scales, by splits in levels that start to co-evolve as soon as they come into existence. Empirically, top-down and bottom-up developments combined, with higher-level structures crystallizing through conquest or cooperation, and local governance delineating itself through intervention of the center.

From the moment a community emerges, and as soon as we have a group of actors making decisions about that community, stories appear that clarify the identity of the community and the position of the early elite (Brockmeier, 2002). Stories about individual identity and community identity tend to appear together. As social complexity increases, more variations appear in this relation between individual identity and the systems of decision-making, which tend to become systems assigning a place to an increasing number of groups, that then strive to acquire a voice in the system (Luhmann, 1990a, 1990b). A staggering variety of forms and development paths marks human history, but one shared feature is that governance has difficulty surviving when there is a tenuous connection between, on the one hand, the structures of power and decision-making and, on the other hand, the beliefs of people, especially their ideas of good governance, of the ideal community, of a life worth living

DOI: 10.4324/9781032696676-1

(Drath & Palus, 1994). Stories are always there; they give meaning to life; and they cannot be ignored forever by those in power. Indeed, authoritarian states did abound from the earliest times, yet even the Egyptian Pharaohs could not do as they pleased if this would entirely contradict what their most humble workers would believe and feel to be a fair deal, an honest reflection of a natural order where the Pharaoh played a naturally elevated role (North et al., 2009). He or she themselves had to stick to the religion believed in by all, even if he was central to it. Ancient Egyptian governance reflected and produced a world where each had a place, a place that had to be respected by all, a world where legitimacy was bestowed upon powers that were expected to maintain it.

For psychoanalysis, people and their communities develop and function on the basis of stories and the sharing of those stories. Those who participate in governance, we will call actors, and they will play a special role in the mediation between individual and collective stories, in the double sense of stories about individuals and the community, and stories cherished by them, playing a role in their self-constitution (Van Assche et al., 2023). Actors can be individuals, groups, and organizations, and the diversity in their nature comes with differences in the mediating role between individual and community, in their positioning between the individual residents and the stories marking the centers of power. What constitutes influence and participation will differ too, and a fine-grained picture of a community will reveal more actors and stories exerting influence.

One can be at the king's table, and the king listens. One can be in the leadership of a craft guild, and represent it at Bruges city council, anno 1243. The head of an American chamber of commerce suggests economic development policies to council. A new village committee invites people to react to regional plans and discuss alternatives. Actors participate in governance, and where direct participation occurs, individual citizens can be actors. In complex societies, direct participation does not suffice and modes of representation develop. Those who participate, however, do not always directly represent groups in society (Pierre & Peters, 2020). With the formation of civil society and industry groups, unions, and religious organizations, with the rise of bureaucracies since the eighteenth century, the differentiation of legal systems from politics, in short, with the increasing complexity of society, the number of actors and lines of influence proliferated, while the connection between some actors and the lifeworld of residents became slight (King & Thornhill, 2003). Lobby groups and unions represent people, however in a limited aspect of their desires. At the same time, in democratic rhetoric the desires of administrative departments are not supposed to count, but naturally they do.

Actors, including administrative ones, have identities, and those identities we will understand in this book as narratively constructed. As actors act, the relation between these stories and what actors actually do, is of great relevance, and psychoanalysis can assist us greatly in understanding it. For psychoanalysis, actions tend to be overdetermined; more than one reason is likely at play. And, not all of those reasons will be readily understood by the actor. Often the forces truly driving decisions would be disavowed if brought into the open (Glynos, 2010; Stavrakakis, 2002). In explaining reality and presenting realistic options for the future, narratives offer attractive tools for simplification. Simplicity is in many ways necessary, but it also creates a variety of problems (Gunder & Hillier, 2009).

In studies of governance, public policy, and public administration and planning in much of the twentieth century and beyond, the drive for clarity and simplicity can be captured under the name of modernism (Allmendinger, 2017; Hillier, 2003; Scott, 2020). One can also consider modernism a metanarrative, a narrative that connects and gives sense to a variety of other narratives (Hampton, 2011). Modernism assumes that reality can be objectively understood and assessed and that actors can be fully aware of their motivations. A psychoanalytic perspective on policy and governance, however, would take issue with all such assumptions. Monolithic understandings of democracy, good governance, community identity, sustainability, and resilient communities look rather unhelpful from a psychoanalytic perspective.

If we understand governance as the taking of collectively binding decisions in and for a community, then the question arises immediately what that community would be. This book will argue that sometimes the community preexists governance, while in other cases governance creates community. Similarly, the nature of the binding of the community and the policy tools used will differ. Institutions for us are those collectively binding rules. We distinguish between policies, plans, and laws, or formal institutions, and, in contrast, informal institutions (Van Assche et al., 2013). In some places, informality will dominate over formality. Where formal institutions are most important, the relative significance of policies, plans, and laws will differ (Ledeneva, 2006). Even where laws are the only institutions that can make something happen, the way it happens can still occur in various ways. Understanding this diversity in forms of communal organization and decision-making requires an understanding of the nested systems of governance, community, and environment. It requires an appreciation of the way the community sees itself and its governance, and a grasp of how governance understands itself. A starting point for a psychoanalytic exploration of governance can be precisely this multiplicity of interpretations that enables a multiplicity of forms of organization (Czarniawska, 2014).

Actors do not exist in the wild and communities do not exist in the abstract. Actors are actors only by becoming part of the governance system, formally or informally, directly or indirectly, and governance only becomes such if there is a community, real or imagined, to which the decisions will apply (Latour, 2004). Decisions thus have the potential to shape the community, transform it in a chosen direction, to strive for the preservation or production of public goods. Governance has the power to create, absorb, and project images of self, environment, and others, of past, present, and future, of feared or desired, and has the power to absorb and craft stories that can shape both the community and the governance system itself (Pottage, 1998; Teisman et al., 2009).

Governance systems reflect and transform social identities; they remember and forget; they can be marked by traumatic histories; they can construct realities that at a certain moment seem to suit them, but they can also impose severe restrictions on the construction of realities and their navigation (Antze & Lambek, 2016). Such restrictions have practical effects, and they can have social, cultural, and political implications. Alternative realities and alternative futures can be thwarted in their construction and dissemination by governance (Žižek, 2008, 2009, 2020). Patterns of inclusion and exclusion in governance and in society are subjected to the powers of governance.

Psychoanalysis and governance

Governance is about choice, about communities as they think and rethink their past, present, and future. It is about subjects and objects, about people and groups who shaped themselves and their environment through stories that will always be contestable and open to new interpretations. Dreams and fears, collective hopes and anxieties, desires acceptable and unacceptable are at the core of governance, even if modernist theories prefer to overlook this messiness (Eberle, 2019; Frosh, 2001, 2013). For psychoanalysis, of course, the mess is the substance. What emerges as a common good, as a goal in governance, does so in a struggle between not only factions with established identities, but also between interpretations of self, world, and environment that are never entirely fixed and never entirely transparent to those pronouncing them (Glynos & Stavrakakis, 2008; Gunder, 2015).

If psychoanalysis is useful to understand governance, it is useful to understand our possibilities for improving the communities and the world we live in. This basic idea will pervade most of the following chapters. Yet, if psychoanalysis is helpful, is it only a matter of better understanding how things work, and from there continuing the normal life of governance? Is it a tool for academics, maybe consultants, to tell richer stories to people who already know what they are doing, or could it offer a mode of self-transformation, a path to real change? If so, who can be regarded as an analyst? What would be the hallmark of a successful analysis? What could trigger the beginning and mark the end of an analysis? And does it even make sense to speak of therapy in the context of communities and their governance? These questions cannot be answered at the outset of our investigation. We will face them in our last chapters. First, we need to develop our understanding of governance, and the features that psychoanalysis can elucidate. What is clear, however, in any democratic context, is that analysis, first of all, has to be self-analysis.

If we forget for the moment the complexities of multilevel governance, we can, for now, assert that outsiders ought not play the role of therapist. The analyst, in other words, can be the community itself. Analysis can take place in governance and beyond. We will need to distinguish roles for therapeutic outsiders under circumscribed conditions. As life cannot be reduced to therapy, self-analysis cannot be the sole purpose of community life. We will delineate the conditions under which more intense forms of self-analysis are appropriate and which roles could be defined. Leadership, it will be argued, is essential, because the initiative to intensify self-reflection and the initiative to embark on a path of community change are linked; they are unlikely to emerge from existing procedures of governance.

The demarcation of our perspective from any form of modernism can be continued along this line. Community therapy cannot be imposed, and even less desirable is it to let outsiders shape the final interpretation of the process or draw practical and moral conclusions (Hinchman & Hinchman, 1997; Hutchison & Bleiker, 2008; Janes, 2016). In all modernist regimes, the tendency to come to objective observations and draw binding conclusions, identified by experts, would be too strong by far. In an analysis of community identity and community desire, the desire and the identity of outside observers cannot be ignored. Moreover, high modernist fantasies of development, of efficient governance, and of correct inclusivity and class structure are not only found in the postwar decades. New Master Signifiers such as sustainability, innovation, resilience, and participation itself tend to be recuperated in a modernist vein, which means that they are embraced by political, administrative, and economic elites more interested in fixing or ignoring identities and desires than in truly understanding them and enabling their self-transformation (Gunder & Hillier, 2009; Voß & Bornemann, 2011). Relying entirely on expert discourse and giving too much weight to higher-level actors in the analysis of a community and in the drawing of conclusions from that analysis, is therefore a risky strategy (Fischer, 2000; Van Assche et al., 2024). Relying entirely on local governance, or participatory decision-making is problematic as well.

As communities can reflect on themselves, decide on possible futures, and come into being through governance, the self-analysis of communities and the drawing of conclusions, must be firmly located in governance. Besides the political and moral problems of analysis imposed by outsiders, there are very strong practical reasons to emphasize the character of analysis as self-analysis. To achieve an effect of a new understanding of *self*, that understanding needs to be recognized in the community itself (Jessop, 2002; Johnston, 2009). It must resonate with existing narratives and values, or establish a connection with them through the process of self-analysis. If apparently neutral facts are thrown at a community in distress, chances are slim that this will have the intended effect.

To be convincing for the community, a new interpretation of reality must find support in the existing understandings in the community. Even if, in the end, the transformation is radical, it needs to start in the discursive and affective landscape of the community. We will argue that a psychoanalytically inspired form of self-analysis is eminently useful when current understandings of self and environment do not produce the public goods aspired to, do not support the kind of community the residents desire. If what is at stake is not only a reinterpretation of self and environment, but also a new community future, then the positioning of those who ask questions or guide the self-analysis will be even more sensitive (Kornberger, 2017; Laclau, 2005; Mintzberg, 2017).

For the community, self-analysis can be helpful beyond better understanding its own desires, as the same analysis can shed a light on the entwining of thinking and organizing, on the adequacy of the governance tools and routines it availed itself of, which it possibly identified with (Alvesson & Spicer, 2012; Margulies, 2022). Thus, while some stories might end positively in an unexpected repositioning of self, others might, more modestly, find their value in an affirmation of identity and desire, and a clarification that problems are, in their case, indeed more technical, more in the domain of organization. Of course, as thinking and organizing do entwine, as stories and policies do keep each other in place, chances are that both sides are involved to a degree.

Layout of the book

Thus, Chapter 2 engages with the themes of memory and identity. How communities look at the future will depend on the way they understand themselves and their past. We distinguish between history, memory, and legacies, three intricately interwoven yet distinct concepts, three concepts relevant for the understanding of governance. Communities, even in their moments of self-conscious reflection, even in specialized roles, are not fully aware of the diversity of legacies from the past that mark the concepts available to understand self and environment, and of the tools to transform. While we leave a full answer of the question of the identity of governing systems for the last chapter, we start with the formulation of preliminary answers in Chapter 2, where we explore the potential of classic psychoanalytic concepts such as Ego, Superego, and collective unconscious for the understanding of regulatory functions in society. Narrative connects the individual, governance, and community. In the construction of past, present, and future, the need for stories is also a need for fiction. Communities and their governance systems cannot function without the clarity and cohesion offered by fiction, yet not all fictions are created equal.

Chapter 4 therefore tackles the theme of reality and reality testing in governance. Maintaining our distance from modernist approaches to governance, we will analyze the diverse mechanisms available to governance systems in the construction of realities, and in the veiling of their constructed, thus contingent nature. Governance systems observe themselves and their environment in imperfect and selective ways. Blind spots always occur, and the patterns of non-observation will have effects on thinking and learning. In a psychoanalytic perspective, reality testing cannot be a process of checking conformity with external realities. Measuring success and failure, understanding problems and solutions all rely on dominant understandings of reality, the good community, and governance itself. Nevertheless, problems do exist, and policy failures are real. Reality testing, therefore, cannot be simply abandoned as a concept and requires reformulation. In the spirit of Mikhail Bakhtin, we come to an understanding that relies neither on monolithic expert knowledge nor on the proliferation of individual opinions and local sentiment. Such polyphonic understanding of reality testing also fits our context-sensitive understanding of governance. Good governance cannot be reduced to one ideal structure, one set of procedures, one balance between levels, between expert and local realities.

We need a detour, in Chapter 3, to reflect more thoroughly on the power of routines, traditions, and standardization in governance. One cannot fully grasp the reality construction and the pattern of blind spots in governance, nor the difficulties in self-transformation if we do not appreciate the power of repetition and the structuring effect of non-thinking. Routines can be powerful tools to support productive fictions, thereby amplifying the power to transform society. Alas, however, routines and procedures can also produce blind spots that are hard to recover from, and they can systematically diminish the power of governance. Processes of naturalization, the hiding of contingency in the construction of realities, are supported by routines that reduce the scrutiny of procedures and decisions. Some stories are more important than others here, as they likely lead to the selection and construction of other narratives, and to encoding in governance routines. When problems occur, and routines do not provide the answer, exceptional circumstances might warrant their suspension. Rethinking routines and reconsidering narratives can appear as a possibility, then, and this opens the door to community strategy and to forms of intense self-reflection that might be called therapy. We reflect on the types of situations where openings appear for the reconfiguration of routines, including shock, conflict, loss of trust, increased uncertainty, and Grand Challenges.

Chapter 5 complicates our story by bringing in affect. Stories have effects that cannot be understood without reference to emotions. Emotional investment in identity narratives brings them to life, gives them effects in governance. Cultural change, shifting narratives can engender new affects, while new affects can colonize older stories and influence their emotional investment, or *cathexis*. Modernist ideologies of evidence-based policy or expert-driven governance attract the most passionate defenders, and these passions are rarely examined. Affect, in other words, is everywhere, and narrative in governance is kept utterly busy in a variety of mediator jobs. Stories enable the expression of ideas and emotions, their connectivity, and they allow for the recognition of ideas and emotions. They build identities that are invested with affect, identities that then produce new stories and guide the emotional investment of others. Hope and desire, anxiety and fear acquire a special relevance in governance, as they are rooted in the past, and affect the orientation of the community toward the future.

Chapter 6 tries to analyze the distance between governance and community, between the stories present in the governance system and what is happening outside. Governance allows the community to represent itself to itself and to draw conclusions regarding desirable futures. In this process, communities can disrupt their imaginary unity, but they can also come to new productive fictions that bring the community into being. Without governance, community identities are not likely to survive for a long time, as consistent effects of those identities, in the form of collectively binding decisions and community change, cannot be observed. Furthermore, as ideas of democracy become normalized, the question of legitimacy appears in a new light: the right to take decisions for the community becomes contested, a diversity of voices along with governance are expected to disrupt an imaginary unity that remains nevertheless craved by the community. Each manifestation of democratic governance thus brings about a new pattern of inclusion and exclusion, of voices, of narratives, of forms of knowledge and preferred tools of coordination. And each pattern of inclusion and exclusion carries the seeds of its own destruction. Governance cannot fully coincide with the community it steers and represents. Differences and friction will always remain. Psychoanalysis reminds us that forgotten options and associated narratives and affects do not simply disappear. They can lead a life in the collective unconscious, where they can trigger effects later on.

Chapter 7 builds on the theme of necessary friction and necessary incompleteness in governance. Friction within governance, friction between governance and the community it represents and steers, comes with patterns of inclusion and exclusion, while those patterns simultaneously reflect the unavoidable incoherence and ambivalence marking governance. Beyond technicalities, the reasons for a particular policy and the effect of that policy are similarly varied and similarly unobserved. The relations between ambivalence and incoherence will be unpacked carefully in this chapter, and this analysis will result in a new understanding of the functioning of and the need for productive frictions in governance. Narratives are needed to gloss over incoherence, incompleteness, frictions, and ambivalence. Suspension of disbelief is an absolute necessity to make governance work. The more ambitious the governance system; the more comprehensive and ambitious its visions for the future, the more important it becomes to operate based on a partly imaginary self-understanding. Legacies of conflict and trauma can become problematic for the cohesion of governance, and for the implementation of strategies to move the community in a desirable direction, as they can make it harder to reflect on ambivalence and inconsistency, and as they might produce symptoms that can be interpreted as the problem itself.

This brings us to Chapter 8, where we introduce the concept of governance paths, borrowed from evolutionary governance theory, indicating the history of co-evolution marking each governance system and bestowing a unique character upon it. It is in a governance path that we can grasp the appearance of new understandings of self and environment, new tools for change. The structures and procedures, memories and narratives that develop over time are both limiting and enabling, in the sense of stabilization of the community and in the sense of shaping its future toward cherished values and desired identities. Psychoanalysis offers insight in both the limiting and the enabling effect of the structures that evolve over time in governance and community, ranging from the dependence on narratives to dependence on others, on a preferred past and future, on selective blindness and productive fictions, on basic scenarios framing the understanding of new situations. Psychoanalysis further illuminates how limitation only becomes a problem when it impedes everyday functioning, or conflicts with deeply held values and aspirations.

In Chapter 9, we face the issue of power, in its entwining with knowledge, in its intricate relations with desire. Desires shift, new discursive connections are produced, new social identities form and reform, and this can reconfigure the relation with perceived needs and policy priorities at any time. Patterns of inclusion and exclusion in governance result from power relations and represent the matrix of possibilities to reshuffle them. The inclusion of actors entails the relevance of their perspective, while the positioning of expert knowledges in procedures automatically bestows power on them. Power/knowledge is analyzed as a driver of governance, which leads us to the question of drives, and here we are forced to observe that governance systems curtail individual and collective drives, but also create conditions where collective drives be unleashed. Leadership, as a source of identification, and in its role of storyteller, can move the community in new directions it might not have imagined before, toward unexpected and inclusive futures. It can also inspire transgression and destruction, and, in its drive to enforce unity, the collapse of the community. We discuss the functions of ethics and the promises of sublimation.

In Chapter 10, we arrive at a point where we can finally look toward the future. How do communities imagine and organize the future, especially when they are hoped to be different from the present? Fantasy structuring existing versions of reality, partly naturalized through governance, can reappear in new stories about a new future. A collective strategy will only work when the process of visioning itself is trusted, when it is accepted as a function of governance. Dreams, ideals, and hopes both mobilize and immobilize, unify and divide; they can stop the process of thinking and they can also sharpen it. Fear, anxiety, obsession, and other modes of self-limitation, conscious or unconscious, will play out in the construction of collective futures. As in previous chapters, we emphasize hope and fear, anxiety and desire, as emotions that emerge in unique patterns in identity formation, and steer the selection, construction, and affective investment of narratives about the future. As governance is more than a narrative world, the legacies of governance also provide communities with limited tools, limited understanding of their potential uses toward a better future. Master narratives on good governance and desirable futures are mutually imbricated and together limit what can be imagined and organized. Unconscious Master Signifiers, disavowed discourse and desire, can further ensnare communities in their attempts to broaden their self-understanding and, from there, what can be constructed as a realistic and desirable future.

Chapter 11, then, addresses situations where this self-limitation is becoming problematic, and where a form of therapy, of self-analysis, might be desirable. Therapy, it will be demonstrated, is not always possible, nor always desirable. We contemplate the nature of therapy, the potential role of specialists and outsiders, of leadership. We also propose styles of leadership that can fit the conditions of communities in need of reinvention, communities where therapy might be appropriate. In considering the beginning and the end, the conditions for and the hoped-for results of therapy, we need to revisit the notions gradually built up in the previous chapters, and remind ourselves what a community is, and what the function of governance could be. We come back to the nature of community identity, which can now be safely presented as both narrative and autopoietic, as spanning the realms of the Imaginary, the Symbolic and the Real. Communities come into being as Voice, and Voice can upgrade governance, can give it access to its higher functions, to the possibility to do more than represent the community to itself. It opens the door to a version of governance that can transform itself and the community toward a future understood as more desirable. Yet, paradoxically, more ambitious modes of governance tend to harden the split between governance and community, while, ultimately futile, attempts to bridge the divide, to erase the gap, result in a collapse of the virtuality of collective strategy.

Concluding

We circle back, thus, to community identity and community future, and reconsider ideas of co-evolution and emergence. If individuals and communities owe their emergence as complex entities to coupling and co-evolution, and if meaning – hence narrative – is the shared medium in that evolution, and if governance is the site where couplings are enacted and reenvisioned, then it is easier to understand why community identity is not always relevant in governance, and why community futures are not always envisioned. Community identity, as an imaginary, as narratively constructed Ego, coexists with an autopoietic identity, with both Real and Symbolic dimensions. While individuals cannot function without Ego, communities do not *need* such to function, yet an autopoietic identity, reflecting the practical functioning and evolution of governance, is always there. Moreover, as soon as ambitions arise, as soon as desire rears its head and repression kicks in, in unconscious forms, with that, both conscious and unconscious Ego functions appear. The subject of the community comprises both aspects of its identity and entails conscious and unconscious dimensions. It is not always there, as, for us, community can unravel, even where formal governance structures remain untouched. Governance, however, can offer an infrastructure for the endurance of collective Ego and subject, and, once in place, for self-reflection and self-transformation. A Lacanian lack, a structuring negativity does need to install itself in the community, for it to recognize itself, and, when self-limitation proves problematic, to transform itself.

The cohesion of the community will vary, the possibilities of governance to reshape community identity and futures will vary, so the phenomena so well known for psychoanalysis will not always emerge, yet the point is that they *can*. Psychoanalytic concepts can help communities grasp both their problems and limitations, as well as their potential to overcome them. Where communities are considering change but feel unable to do so, a psychoanalytic perspective can greatly enhance their capacity to reframe their own condition.

As we argued throughout this introduction, this can take a twofold approach, first, by elucidating how community identities emerge and transform through governance, and second, by dispelling common mythologies on good governance, inspired by the Loch Ness monster of undying modernism. These arguments, however, can be made much more compellingly after winding our way through the narrative of the following chapters, so we politely nudge our readers to move along and look, with us, at the puzzling matter of history, memory, and legacy in governance, the topic of our second chapter.

References

Allmendinger, P. (2017). *Planning Theory* (3rd ed., p. 346). Macmillan Education UK.

Alvesson, M., & Spicer, A. (2012). A stupidity-based theory of organizations. *Journal of Management Studies, 49*(7), 1194–1220.

Anderson, B. (American C. of L. S.) (2006). *Imagined Communities: Reflections on the Origin and Spread of Nationalism* (3rd ed., p. 240). Verso.

Antze, P., & Lambek, M. (Eds.). (2016). *Tense Past* (0 ed.). Routledge.

Brockmeier, J. (2002). Remembering and forgetting: Narrative as cultural memory. *Culture & Psychology, 8*(1), 15–43.

Czarniawska, B. (2014). *A Theory of Organizing*. Edward Elgar Publishing.

Drath, W. H., & Palus, C. J. (1994). *Making Common Sense: Leadership as Meaning-making in a Community of Practice*. Center for Creative Leadership.

Eberle, J. (2019). Narrative, desire, ontological security, transgression: Fantasy as a factor in international politics. *Journal of International Relations and Development, 22*(1), 243–268.

Fischer, F. (2000). *Citizens, Experts, and the Environment: The Politics of Local Knowledge*. Duke University Press.

Frosh, S. (2001). On reason, discourse, and fantasy. *American Imago, 58*(3), 627–647.

Frosh, S. (2013). Psychoanalysis, colonialism, racism. *Journal of Theoretical and Philosophical Psychology, 33*(3), 141–154.

Glynos, J. (2010). *?Lacan at Work?* In C. Cederstrom & C. Hoedemaeker, (Eds.), Lacan and Organization (pp. 13–58). MayFly Books.

Glynos, J., & Stavrakakis, Y. (2008). Lacan and political subjectivity: Fantasy and enjoyment in psychoanalysis and political theory. *Subjectivity, 24*(1), 256–274.

Gunder, M. (2015). The Role of Fantasy in Public Policy Formation. In R. Beunen, K. Van Assche, & M. Duineveld (Eds.), *Evolutionary Governance Theory: Theory and Applications* (pp. 143–154). Springer International Publishing.

Gunder, M., & Hillier, J. (2009). *Planning in Ten Words or Less: A Lacanian Entanglement with Spatial Planning*. Ashgate Publishing, Ltd.

Hampton, G. (2011). Narrative policy analysis and the use of the meta-narrative in participatory policy development within higher education. *Higher Education Policy*, *24*(3), 347–358.

Hillier, J. (2003). `Agon'izing over consensus: Why Habermasian ideals cannot be `real'. *Planning Theory*, *2*(1), 37–59.

Hinchman, L. P., & Hinchman, S. K. (1997). *Memory, Identity, Community: The Idea of Narrative in the Human Sciences*. State University of New York Press.

Hutchison, E., & Bleiker, R. (2008). Emotional reconciliation: Reconstituting identity and community after trauma. *European Journal of Social Theory*, *11*(3), 385–403. https://doi.org/10.1177/1368431008092569

Janes, J. E. (2016). Democratic encounters? Epistemic privilege, power, and community-based participatory action research. *Action Research*, *14*(1), 72–87. https://doi.org/10.1177/1476750315579129

Jessop, B. (2002). Governance and Meta-governance in the Face of Complexity: On the Roles of Requisite Variety, Reflexive Observation, and Romantic Irony in Participatory Governance. In H. Heinelt, P. Getimis, G. Kafkalas, R. Smith, & E. Swyngedouw (Eds.), *Participatory Governance in Multi-Level Context: Concepts and Experience* (pp. 33–58). VS Verlag für Sozialwissenschaften. https://doi.org/10.1007/978-3-663-11005-7_2

Johnston, A. (2009). *Badiou, Zizek, and Political Transformations: The Cadence of Change*. Northwestern University Press.

King, M., & Thornhill, E. (2003). *Niklas Luhmann's Theory of Politics and Law*. Palgrave Macmillan.

Kornberger, M. (2017). The values of strategy: Valuation practices, rivalry and strategic agency. *Organization Studies*, *38*(12), 1753–1773. https://doi.org/10.1177/01708 40616685365

Laclau, E. (2005). *On Populist Reason*. Verso.

Latour, B. (2004). *Politics of Nature: How to Bring the Sciences into Democracy*. Harvard University Press. https://doi.org/10.4159/9780674039964

Ledeneva, A. V. (2006). *How Russia Really Works: The Informal Practices That Shaped Post-Soviet Politics and Business*. Cornell University Press.

Luhmann, N. (1990a). *Essays on Self-Reference*. Columbia University Press.

Luhmann, N. (1990b). *Political Theory in the Welfare State*. De Gruyter.

Margulies, J. (2022). A political ecology of desire: Between extinction, anxiety, and flourishing. *Environmental Humanities*, *14*(2), 241–264. https://doi.org/10.1215/22011919-9712357

Mintzberg, H. (2017). *Crafting Strategy*. Routledge.

North, D. C., Wallis, J. J., & Weingast, B. R. (2009). *Violence and Social Orders: A Conceptual Framework for Interpreting Recorded Human History*. Cambridge University Press.

Pierre, J., & Peters, G. B. (2020). *Governance, Politics and the State*. Bloomsbury Publishing.

Pottage, A. (1998). Power as an art of contingency: Luhmann, Deleuze, Foucault. *Economy and Society*, *27*(1), 1–27.

Scott, J. C. (2017). *Against the Grain: A Deep History of the Earliest States*. Yale University Press.

Scott, J. C. (2020). *Seeing Like a State: How Certain Schemes to Improve the Human Condition Have Failed*. Yale University Press.

Stavrakakis, Y. (2002). *Lacan and the Political*. Routledge.

Teisman, G., van Buuren, A., & Gerrits, L. M. (2009). *Managing Complex Governance Systems*. Routledge.

Van Assche, K., Beunen, R., & Duineveld, M. (2013). *Evolutionary Governance Theory: An Introduction*. Springer.

Van Assche, K., Beunen, R., & Gruezmacher, M. (2024). *Strategy for Sustainability Transitions: Governance, Community and Environment*. Edward Elgar Publishing.

Van Assche, K., Gruezmacher, M., Marais, L., & Perez-Sindin, X. (2023). *Resource Communities: Past Legacies and Future Pathways*. Taylor & Francis. https://books.google.ca/books?id=4EHUEAAAQBAJ

Voß, J.-P., & Bornemann, B. (2011). The politics of reflexive governance: Challenges for designing adaptive management and transition management. *Ecology and Society, 16*(2), 9. [online] URL: www.ecologyandsociety.org-9

Žižek, S. (2008). *Violence*. Picador.

Žižek, S. (2009). *In Defense of Lost Causes*. Verso. http://public.ebookcentral.proquest.com/choice/publicfullrecord.aspx?p=5176960

Žižek, S. (2020). *The Plague of Fantasies*. Verso Books.

2 Memory and identity

Memory, history, legacy

Stories require memory, and memory in governance is not to be described as one thing, but as a set of entwined, sometimes mutually hampering functions. Memory is always selective, as remembering everything would result in information overload. Unstructured complexity would be of little use for navigating life and for the maintenance of identity (Antze & Lambek, 2016). Memory builds stories and derives its selectivity partly from other stories. Forgetting is hence not simply the other side of the coin, as that which is forgotten is not only that which is not selected, but also that which is deliberately forgotten. What is forgotten might be repressed, inaccessible for the collective, but it might also lead a lingering existence in corners of the community and its governance system (Frosh, 2001, 2013).

Memory in governance systems is not a copy of memories entering the system through the selectivity of elections. Nor can it be grasped as the sum of all memories belonging to individuals elected as officials, active in administration, or otherwise connected to collective decision-making. Each organization involved can develop its own identity, and hence memory, and the set of organizations we call administration, as well as the governance system as a whole, will be marked, over time, by unique identities and memories (Esposito, 2008; Luhmann, 2018). The way things are remembered and forgotten is guided by stories about self, environment, about the relation between governance and community.

Memory in governance is supported by infrastructures. At the same time, the memory infrastructure of governance can shape the realities in the communities. It is in governance that collective memory can be shaped and most effectively preserved, that values can be translated into policies, that past, present, and future can be connected (Van Assche et al., 2009). Administration, its expertise, its relative stability, can play a role here, as well as archives inside and outside government, libraries, museums, monuments, heritage policies, events, and research organizations (Brockmeier, 2002). In complex governance systems, a diversity of memories will be preserved, leaving

DOI: 10.4324/9781032696676-2

space for alternative interpretations of reality, and diverse policy responses (Wagenaar, 2004, 2015).

History is a form and function of social memory, as in a standardized and officially sanctioned version of events, usually structured in narratives tied to the mythologies of the nation-state or of smaller communities (Anderson, 2006; Scott, 1998). History is a version of what happened that can be shared within the community according to the powers that be, and that can serve to increase cohesion. Alternative histories can emerge if a community of readers and experts develops, giving its own stamp of approval, disputing the dominant identity narrative, often promoted by the nation-state. Schools, universities, and museums reproduce history, reflect on it, within parameters that maintain a distinction with mere stories. Facts are constructed within interpretive frames which rely on other stories, identities, and methodological assumptions that all require legitimation in and by governance and a community (Foucault, 1968, 2012; Yanow, 2000). History, as memory, is always rewritten, under the influence of changing culture, notions of expertise, shifting relations between subgroups in society, and political discussions.

Legacies are broader than history and memory. The reproduction of governance, with new institutions emerging out of interactions between sitting actors and their use of existing institutions, is subject to the co-evolution of actors and institutions, power, and knowledge, meaning that each step in governance evolution is constrained by a variety of legacies, observed and unobserved (Jessop, 2002; Teisman et al., 2009; Van Assche et al., 2013). History and memory are observable legacies, providing orientation for the future. Even in their cases, the selectivities at work are not entirely observable. Schools in a nation-state are only partially aware of the frames guiding their teaching of history, of the importance of old versions of the nation-state.

Legacies can be *organizational, material, and cognitive* (Van Assche et al., 2023, 2024). Some legacies are relevant for the entire community; others leave their deepest traces in governance. Governance develops through building on existing patterns of governance; it survives due to strong organizational legacies (Luhmann, 1989, 1990). Forms of organization emerging outside the sphere of governance can affect it by providing models of organizing, or through the consolidation of voices that then link to governance, or, alternatively, that offer resistance. Within governance, stories encoded in policy tools or other institutions, as well as patterns of expertise, have limiting effects on the learning options in governance; they constitute cognitive legacies (Dunlop & Radaelli, 2020; Voß & Bornemann, 2011). Such encoded stories can shape the reality as perceived by governance actors, and depending on the coupling between governance and community, on what counts as reality for the community.

Legacies can be observable at first, then slowly move out of the scope of observation, while remaining active (Duit & Galaz, 2008). In the other direction, unacknowledged or repressed legacies might slowly become visible, as when public discourse shifts, and what looked normal and natural before now starts to look contingent and questionable (Wagenaar, 2015). Transparency has limits, as the narratives which structure the identities of individuals, organizations, and community are never entirely observable from within, and as these narratives further structure observation in a manner that is not entirely conscious (Valentinov et al., 2018). The interplay between actors in governance, between governance and community can help elucidate blind spots, and bring to the surface what was not discussed before, but in this interplay, in addition to clarifications, new repressions and blind spots will appear. The multiplication of perspectives enables reflexivity in governance, and hence increased transparency, but the nature of each clarifying discourse and the nature of discourses interpreting each other, make that the result cannot be a transparent universe and a community transparent to itself (Hillier & Gunder, 2003). The mechanisms of democracy, including the possibility of vibrant public discourse, cannot abolish the opacity and ambivalence of the individual. Rather, the way both individuals and communities build themselves through narrative, and the role of the unconscious in identity formation, make for a collective that displays not fewer but more numerous forms and functions of intransparency (Glynos, 2010). The Freudian iceberg did not melt through mutual clarification.

Narratives and Master Signifiers

Some narratives are more important than others, and some *signifiers* are more important than others. Following Jacques Lacan, we speak of *Master Signifiers*, signifiers that are open to interpretation yet able to seemingly pin down meaning, to accommodate internal difference while enabling the continuation of discourse and decision-making (Hook, 2017; Kooij et al., 2014). The meaning and relationality of these and other psychoanalytic concepts will be enriched in this book. Social and organizational identity requires productive fictions, signs, and stories that can structure other signs and stories (Van Assche et al., 2020). Master Signifiers are not always conscious, as they can be repressed or disavowed. "Sustainability" can be a Master Signifier, but so can be "white supremacy." Master narratives are narratives that underpin, structure, stabilize, guide the production of other narratives, in the discursive figurations that create the realities for community and governance system. They can be assembled around Master Signifiers, which can be the driving force for the narrative structuring, but in our view, it can also work the other way around, with Master Signifiers coming into being through the gradual formation, linkage, and hardening of narratives.

Stories of special significance for the understanding of governance are ideologies, big stories that purport to explain big issues, sometimes with an emphasis on guidance through reality, sometimes constructing reality with only hints regarding its navigation (Wardle, 2016; Žižek, 2019). Some ideologies are more normative than others, more rigid, more comprehensive. Some ideologies focus on the inclusion of diverse other narratives or subjectivities in governance, while others can consciously limit its meaning, function, and scope (Swyngedouw, 2022). Ideologies can be connected to other master narratives, and these connections can emerge in histories of which only few people might be aware. The religious background of a set of local laws, administrative practices, might elude most people; an idea of freedom might be tightly coupled to a tightly circumscribed version of good governance.

Identity stories, which require and produce memory, and can enable the navigation through complex realities, can function as master narratives. They can be among the most significant legacies in governance systems, especially if the identity narrative has clear implications for governance, and if it associates with forms of governance deemed desirable and natural (Van Assche et al., 2023). Even where they claim a deep history, in practice they are continuously renegotiated. Distinctions will be invoked and generated, yet those distinctions do not provide a stable ground either. Gender identities do not only shift over time; they also shift in relevance for governance, and the same applies to class distinctions, ethnic boundaries, and the associated identity narratives (Skeggs, 2013; Snyder, 2015). Identity narratives are of course not the only stories structuring governance. We mentioned ideologies, and we can add stories of good governance, not always coupled to ideologies, and other stories that serve to stabilize governance or society (Delanty, 2016; Vaara et al., 2016). Governance can make interactions more predictable, and realities more acceptable, trusted, and shared. Stories of success, procedures of defining problems and solutions, stratagems for hardening realities (see Chapter 4), methods to limit doubt, to pre-structure criticism and variation in perspectives, to manage conflict, to instill pride, all help in the mutual stabilization of governance and community.

Ego, superego, subject

Drawing on Sigmund Freud and Lacan helps us see the Ego, as a productive, organizing fiction, as a grounding fantasy, in the realm of the imaginary (Fink, 1999). Not every group, we argue, is a collective marked by an Ego, however. The following passages and chapters develop this idea, in its relation to the potential roles of governance. Identity as Ego, as narrative construct guiding the interpretation of self, simultaneously co-constructing that self, does not always appear (cf. Vanheule, 2016). One can draw parallels with organizations, where the existence of a name, a building, even roles, does not indicate there is a truly functioning organization (Luhmann, 1995). Ego, therefore, must be observed empirically. Self-descriptions cannot be relied upon, as there are most likely various rhetorical self-images bounced around in governance and society, without their having much effect on the actual self-image.

The role of identity narratives in the couplings between the individual, the governance system, and society differs widely (Healey, 2004). In other words, the meaning of society for the individual, the meaning of governance for the individual and society, and the role of governance in shaping society vary over time and space. We take meaning to be the shared medium for individuals and social systems, and narratives of identity *can* structure the production of meaning at all levels, potentially establishing couplings between levels, allowing for an influence of individual Ego on governance and community but also for the formation of a collective Ego. As governance is by definition organizing, as it can never be reduced to the mere act of storytelling, or to the mere collection or cohabitation of stories, the institutionalized process of organizing the community and taking decisions with that aim, takes on a role in expressing but also enabling the collective Ego (Stavrakakis, 2002). Realities become reality through organization, while organization expresses and promotes a version of reality (Czarniawska, 2013).

Speaking of the collective Superego in governance becomes possible in this perspective. Superego for us is not a higher function of the collective, protecting us from sliding into immorality. Superego functions can appear in various places in governance and public discourse, and can drive toward transgression as much as rule-following (Žižek, 2009). Aggression can mark manifestations of Superego, in the urge to impose rules, procedures, rigid norms of behavior, in the ease of forgetting boundaries when pursuing collective goals, enforcing rules, following leaders, imposing unifying stories. A direct connection to the collective *id*, to the basest collective drives can linger (Flescher, 1949; McGowan, 2019). The existence of a collective, and a governance system capable of binding that collective by means of rules, makes it possible for drives to manifest themselves, to become more impactful, to appear as if deriving from high moral standards. Unconscious Master Signifiers can be at work.

The Ego, in our Lacanian-inspired analysis, is a narrative construct, belonging to the order of the Imaginary (Driver, 2009; Glynos, 2011). The collective Ego cannot be maintained without governance, yet the functioning of governance introduces new opacities and a new form of unicity. We will speak in later chapters of the autopoietic identity, reflecting self-reproduction in a manner not transparent to the Ego or the collective. In the presence of narrative and autopoietic identity, one can speak of a collective subject. Throughout this book, we will build an argument around the centrality of governance for the collective subject, a subject that, in Lacanian fashion, is nevertheless immediately alienated from itself by the split introduced in the community, by a governance system in which the community can see itself and through which it can reshape itself.

The unconscious

Similar to the idea of a collective subject, our tailoring of the concept of the collective unconscious needs to take place throughout the book, not only in this early chapter. What can we say now? First, that we do not understand, as Carl Jung, the collective unconscious as the repertoire of symbols common to humanity, nor the extension of this treasure trove idea to a collection of narratives or narrative structures, of mythologies perhaps, which express shared values, aspirations, fears, or shared structures of meaning-making. As for us, meaning is the medium for the formation of individual and social identities, the collective unconscious is a place of meaning production, not of unarticulated desires, drives, or instincts (Desmet, 2022; Hook & Vanheule, 2016). Those meanings might not be immediately accessible to us, but they are there. They might take on different forms and disguises, but they are structured and, in principle, decipherable.

The narratives we build our worlds by require collectives, as we cannot survive as individuals, in a practical sense and in a psychological sense. We are linguistic beings, and a language of one does not exist. Communities are required in order for us to thrive, to develop our own individuality, to reach our potential, to understand ourselves in a way that makes it even possible to grasp that potential: families, cultural groups, nation-states, among others (Kapoor, 2005, 2015). Meaning is the medium for the formation of identities at all levels, and such a process is always one of nested and overlapping narratives, of co-evolving scales (Foucault, 2012; Luhmann, 1995). Yet, as soon as something defines itself by means of narrative, by signifiers, by means of linguistic positioning in a group, alternative understandings of reality, alternative constructions of self are suppressed, and a matrix of potential identifications starts to form. A collective unconscious develops and, with that, forms of collective repression (Soler, 2015).

One might counter that, even if this is true, governance, as a process of rule-making and implementation, as a process of continuous expression and reshaping of the collective, serves to make the unconscious conscious, by reflecting on what might be driving the community, and to manage collective affects in rational ways. No unconscious would be left if governance did its job, rationally understood, well enough. We would reply that the functioning of governance leads to new repressions all the time, that, indeed, it requires continuous repression in order to live with itself and its limitations (McGowan, 2017; Žižek, 1991, 1992). Governance, both expressing and producing realities, is routinely repressing and acknowledging meanings and requires *fiction*. We speak of productive fictions that enable the community to support other stories, and, in the process, move toward collective goals, in turn structured by fantasy (Catlaw & Marshall, 2018; Eberle, 2019; Gunder, 2005).

A primary form of repression in governance is the nonrecognition of these fictions as fictions. What we do not want to know is how we as a community have limited control over our fate, over internal cohesion, and limited understanding of the real possibilities and threats facing us. We do not want to know that we operate on the basis of fictions that keep each other in place, of ascriptions of necessity and causality that are entangled with fantasy, fictions shaped by the contingent histories of our communities, its systems of coordination and its systems of communication (Žižek, 2002, 2009). This is entirely understandable, as individuals require collectives and their systems to thrive, and that includes the sense of certainty that is guaranteed by those systems. Governance systems help encode realities and *everyone* is dependent on those realities, that is, individuals, communities, and the governance system itself (Ruti, 2012). Realities are not entirely produced by governance but in complex societies; the sharing of realities is a necessity to function for all. The possibility of looking forward as a community (see Chapter 10) hinges on the presence of a present that looks real enough, stable enough (Kaplan & Orlikowski, 2013; Rasche, 2008; Ziai, 2004).

A second, intimately related, form of early repression, not in a historical but a conceptual sense, is that of the origin of the community itself. Communities might know their history, and governance systems can have archives, yet at the same time there is a tendency to extend the present to the past, not to investigate the contingent nature of the community itself, of the perspective through which the community forms itself (Alpert & Goren, 2016; Berger, 1988; Stapleton & Wilson, 2017). Indeed, every community, as every individual, in psychoanalytic perspective, is a perspective on reality that is the product of a contingent series of encounters with that reality, a perspective that allows for certain responses and not others. Hence, selective blindness, and hence founding mythologies, of which recent versions include rationalizing versions of governance as providing services and of community as the group receiving services.

The collective unconscious cannot be fully understood by reference to those primary forms of repression. We encountered Master Signifiers and narratives that elude observation, and our image of the collective unconscious will be enriched after we grasp more of the governance mechanisms which can move ideas and affect in and out of consciousness. We posit unambiguously, though, that all defense mechanisms available to the individual are available to the collective, and that careful observation and reconstruction of processes of governance will bring this to the surface.

Narrative unity and Voice

In governance, decisions must be taken, translated into policies, plans, and laws, into types of formal institutions. Inside and outside governance, a wealth of informal institutions resides, either as parallel coordination mechanisms or as part of the culture in which the governance systems are embedded (Ledeneva, 2018). Narratives in governance can derive from the world of informal institutions outside governance, or not. They can derive from informal institutions inside governance, or not. They certainly might be associated with existing formal institutions. Rephrasing and summarizing these ideas, one can distinguish official narratives, connected to formal institutions, to forms of expertise and versions of reality sanctioned by the governance system, from all other narratives, sometimes leftovers of old or supposedly rejected narratives, sometimes new ones seeping in from the community, without official status – yet (Hillier, 2000; Pusca, 2007).

Official narratives, in democracies, are composite, meaning that they assemble elements of different origin. Their production is expected to be a process of deliberation in political circles, bricolage in administration, discussion, and consultation in other circles (Sehring, 2009). Composite character and assemblage processes do not necessarily bring about a unified Voice. An official narrative might not have formed, resulting in non-coordination, or in a technocratic version of decision-making that purports not to need narrative. Where a unified narrative does appear, connected to institutions to give it impact, this impact can be more limited when the narrative is not recognized as Voice (Bryant, 2016; Palestrino, 2022) . This can be the case because the assemblage was not effective, or because the official narrative is at odds with disavowed master narratives or signifiers. It can be more difficult when internal disagreements cannot be glossed over, or when there is an implicit resistance against speaking with one Voice.

The recognition of Voice is only one step in the process of coordination. Voice can trigger resistance, just as non-Voice can trigger nonrecognition of direction, ambition, or real solution. Recognition of Voice, of a speaking collective can help in the recognition of a governance entity as an actor, while, conversely, the acceptance of an actor supports the recognition of Voice. Formally or academically designated actors are not always real, in terms of actual participation, in terms of recognized cohesion. The appearance of Voice is a key moment in governance, in the coalescing of actors as actors, in the formation of governance identity. Affective investment in governance narratives can find significant support once Voice is recognized.

Composite narratives can still manifest Voice, and a variety of factors are at play here. Legitimacy is one, as are leadership, narrative traditions and styles (Alvesson & Einola, 2019; Gardner et al., 2021). Attitudes with regard to governance, inspired by narratives of governance, influence the recognition and functioning of Voice. As Voice coincides with an ascription of identity, with the possibility of tight coupling between the individual and the community, these attitudes have implications for the functioning of the collective Ego and Superego, the conscious and the unconscious. If whatever is uttered in a governance arena is immediately interpreted as personal, if any policy emerging from that arena is right away rejected as a smoke screen for what is really happening, or as directly benefiting the dominant factions, then chances are that Voice will not be recognized, or, alas, will be a Voice which does not represent the collective and the collective good (Ledeneva, 2006).

Stories that are the result of decision-making, the output so to speak, are almost always affected by the process, in content, and in the interpretation by the community. The more people know about the process, the more complex interpretations are possible, the more alternatives become visible. Almost always, except in the most authoritarian societies, or the smallest and most traditionalist villages, stories result from negotiation and competition and remain in a precarious relation with existing narratives inside and outside governance. The process is not forgotten; what was felt is still felt, and diverging desires and rivalries are not erased. Voice remains fragile, because governance is trying to achieve something, beyond the telling of a story, and because there are other stories present, other interests (Pohl & Swyngedouw, 2023).

Friction between the exigencies of the governance system and the sensibilities of the community render Voice tenuous. Policies never achieve exactly what they are set out to do, governance cannot function precisely according to its public self-descriptions, and trust in governance is not unconditional; hence, the attitude of audiences in the community cannot be expected to be entirely welcoming (Luhmann, 1989, 1990). Story, policy, and the larger world never seamlessly fit together and unproblematically transform each other; hence, the appearance of Voice must be considered an evolutionary achievement, an unlikely outcome that became less unlikely because of the co-evolution of actors and institutions in governance, and more fundamentally because of the constitutive split bringing governance and community into being. Recognizing Voice requires a suspension of disbelief, a recognition of self in governance, and an organization of self through governance (Frosh, 2001; Wardle, 2016). Yet, disbelief is never far around the corner. What is recognized as Voice comes out of this imperfect world, these half-understood relations between governance, community, and environment.

Political and administrative competition, local and expert knowledges that never fully align generate centripetal forces, further conditioning the emergence of Voice and, from there, Ego. The collective Ego remains subjected to these forces. The appearance of Voice is a sign that the Ego is real, while an Ego already believed in, makes it easier for a Voice to be forged out of dissimilar elements, for it to be recognized as such. If the Voice is merely the voice of authority, it will cease to function as Voice, as people tend to need more than coercion to recognize Voice or be subsumed by it. Voice can disappear by losing cohesion. Its narrative spell can be broken, its persuasive power can dampen, when the discursive and affective elements come apart (Žižek, 2019). Its imputed connection to a shared identity can vanish in the blink of an eye. The collapse of Voice and the shift in position vis-à-vis that Voice can occur simultaneously, as there is little reason to believe something if there is no recognizable someone saying it. And there is no real reason to see someone if there is no clear Voice.

As Voice is a sign of identity and vice versa, Voice relies on memory for recognition and maintenance. The functioning of memory introduced earlier in this chapter thus shapes the paths through which Voice can form and become persuasive. Memory modulates the effects of stories of stability, unity, harmony, and success (Beunen et al., 2013; Gunder, 2006). Diversity of perspectives in community and governance is a factor as well, as the awareness of alternative explanations renders something truly stabilizing, even a source of identification with the regime, or rather an object of mockery. Moreover, if existing memory and diversity are not recognized, not validated in the stories coming out of decision-making arenas, mockery can be just the beginning of a more protracted negative response, which can involve resistance, backlash, and counterstrategy (Van Assche et al., 2024).

A story can become Voice if performed well. The act of storytelling we call performance, and performance in governance contribute to the persuasive character of the story, its perception as cohesive, as representing fair process and linking to shared values and narratives. Identities require performance, too, to make other narratives and performances more convincing (Bal, 2002; Howlett & Ramesh, 2014). External validation of identity, via performance, can be a sine qua non for the actor in governance to keep believing in the realities so vigorously defended in the daily grind of politics and administration. Performances of success help to maintain identity, to reinforce master narratives, and to keep governance on a path, chosen or not. Productive fictions can extend to the past, to the moment where supposedly an identity was chosen, an ideology objectively recognized as superior.

Individual, governance, community

Identifications with the narratives of governance, identification of the community with the governance system itself, tend to be unstable, in the fashion of the Freudian hysteric. Overly high expectations for the community, structured by unacknowledged desire, by fantasies giving free rein to the affects otherwise hidden under expert or bureaucratic discourse, can be maintained for a long time, as collective identities can support them, and performances of success can be made immune to critique (Panizza & Stavrakakis, 2020). Hysteric oscillation between identification and alienation, between overly positive and negative judgments regarding governance, appears when both poles are attractive, when a stable identity did not form, when trauma occurred, when new communities emerge, when the world at large makes it known that alternative value systems exist.

The positioning of individuals versus the governance system and the stories it tells can thus be varied and is often ambivalent. People might criticize a regime, but identify with some of its features, or with its grounding mythology. They might attack a particular policy, yet in that attack use, unconsciously, many policy narratives and Master Signifiers stemming from the regime, or, in less authoritarian places, from the dominant culture (Foucault, 2012; Penz & Sauer, 2019). The governance system itself can thus be identified with the community, or opposed to it, through a set of strategic simplifications, and yet, this pattern is never stable, never convincing for a long time, as the game of identifications is never finished. Collective identities can talk back; they can re-relate to other collectives, to other organizations, if there is Voice, if there is Ego. Individuals, always able to draw on different discourses, can shift position, can be unaware of the complexity of their own identifications. Thus, criticizing governance can be latent self-critique or a way of amplifying self-critique. It can be inviting punishment, public atonement, or resocialization.

Ideas of the model citizen and the model community, as produced by governance discourse, or by opposing discourse, can similarly be used in different ways. Not only can people outside governance believe in these ideals, but they can also relate them to reality and to their self-understanding through diverse pathways not often explored. Why are we so upset about this small policy change? Why does this other idea inspire such high hopes? We come back to this issue in later chapters, when discussing affect, desire, and reality testing.

Concluding

Communities evolve. They develop, and do so by guiding systems of self-steering, of self-representation that require memory. The past leaves a variety of traces, however, ranging from officially accepted and promoted histories, to cultivated memories, and to a wide range of cognitive, material, and organizational legacies. As thinking and organizing co-evolve, and governance systems develop with increasing capacities and ambitions, those systems can assume the capacity of Voice, a capacity that does not result from formulaic institutional design or democratic practice, and that has to be described as an unlikely evolutionary achievement.

The formation of community out of a group, of a collective narrative identity we call the collective Ego, has to be associated with the appearance of Voice, and the mediating and supportive effect of governance systems in such a process. Voice is a sign of Ego, and Ego enables Voice. Because the medium of meaning is shared by the individual and the collective, and since identifications play out in language and in group, a history of Ego development is unaidedly accompanied by a history of repressed identifications, of unacknowledged affect and desire, of silent Master Signifiers. In other words, a collective unconscious will form that follows history as a shadow, yet one that, we will see, embodies resilience, identity, and creativity. Neither Ego nor unconscious is in continuous existence, as they rest on the tenuous possibility of Voice, and on the flaring and taming of tensions between the community and the governance system that is supposed to represent and shape it.

Later chapters will develop these grounding notions and unpack the implications for governance and its possibilities for self-understanding and self-improvement. A few things are astoundingly clear already, which is not that surprising for psychoanalysis but is rather puzzling for mainstream theories of politics and administration. A community does not always exist, and governance, its memory functions, its stories and emerging identities, and its tools of coordination can assist in bringing it into existence and stabilizing that existence, making the occurrence of community, of Voice, the functioning of Ego, more likely, and creating conditions for self-transformation rather than collapse. The next chapter analyzes such mechanisms of self-stabilization. Next, we can say that the development of a conscious comes with the evolution of an unconscious, and that legacies from the past render the unconscious opaque for governance. Looking forward, constructing desirable futures and organizing for them, rests on the appearance of both the conscious and the unconscious, as without Ego, no persuasive narrative futures will be built, and without the unconscious, desire will not keep those futures persuasive for long.

References

Alpert, J. L., & Goren, E. R. (2016). *Psychoanalysis, Trauma, and Community: History and Contemporary Reappraisals.* Taylor & Francis.

Alvesson, M., & Einola, K. (2019). Warning for excessive positivity: Authentic leadership and other traps in leadership studies. *The Leadership Quarterly, 30*(4), 383–395. https://doi.org/10.1016/j.leaqua.2019.04.001

Anderson, B. (American C. of L. S.) (2006). *Imagined Communities: Reflections on the Origin and Spread of Nationalism* (3rd ed., p. 240). Verso. https://books.google.ca/books?id=nQ9jXXJV-vgC

Antze, P., & Lambek, M. (Eds.). (2016). *Tense Past* (0 ed.). Routledge. https://doi.org/10.4324/9781315022222

Bal, M. (2002). *Travelling Concepts in the Humanities: A Rough Guide.* University of Toronto Press.

Berger, B. M. (1988). Disenchanting the concept of community. *Society, 25*(6), 50–52. https://doi.org/10.1007/BF02695775

Beunen, R., Van Assche, K., & Duineveld, M. (2013). Performing failure in conservation policy: The implementation of European Union directives in the Netherlands. *Land Use Policy, 31*, 280–288. https://doi.org/10.1016/j.landusepol.2012.07.009

Brockmeier, J. (2002). Remembering and forgetting: Narrative as cultural memory. *Culture & Psychology, 8*(1), 15–43. https://doi.org/10.1177/1354067X0281002

Bryant, L. R. (2016). Symptomal knots and evental ruptures: Žižek, Badiou, and discerning the indiscernible. *International Journal of Žižek Studies, 1*(2), 33–63.

Catlaw, T. J., & Marshall, G. S. (2018). Enjoy your work! The fantasy of the neoliberal workplace and its consequences for the entrepreneurial subject. *Administrative Theory & Praxis, 40*(2), 99–118. https://doi.org/10.1080/10841806.2018.1454241

Czarniawska, B. (2013). *A City Reframed: Managing Warsaw in the 1990's.* Routledge. https://doi.org/10.4324/9781315080048

Delanty, G. (2016). Reinventing Community and Citizenship in the Global Era: A Critique of the Communitarian Concept of Community. In E. A. Christodoulidis (Ed.), *Communitarianism and Citizenship* (Vol. 33, p. 248). Routledge.

Desmet, M. (2022). *The Psychology of Totalitarianism.* Chelsea Green Publishing.

Driver, M. (2009). Struggling with lack: A Lacanian perspective on organizational identity. *Organization Studies, 30*(1), 55–72. https://doi.org/10.1177/0170840608100516

Duit, A., & Galaz, V. (2008). Governance and complexity—Emerging issues for governance theory. *Governance, 21*(3), 311–335.

Dunlop, C., & Radaelli, C. (2020). The Lessons of Policy Learning: Types, Triggers, Hindrances and Pathologies. In G. Capano & M. Howlett (Eds.), *A Modern Guide to Public Policy* (pp. 222–241). Edward Elgar Publishing. https://doi.org/10.4337/9781789904987.00024

Eberle, J. (2019). Narrative, desire, ontological security, transgression: Fantasy as a factor in international politics. *Journal of International Relations and Development, 22*(1), 243–268. https://doi.org/10.1057/s41268-017-0104-2

Esposito, E. (2008). Social Forgetting: A Systems-Theory Approach. In A. Erll, & A. Nünning (Eds.), *Cultural Memory Studies: An International and Interdisciplinary Handbook* (pp. 181–190). Walter de Gruyter.

Fink, B. (1999). *A Clinical Introduction to Lacanian Psychoanalysis: Theory and Technique*. Harvard University Press.

Flescher, J. (1949). Political life and super-ego regression. *The Psychoanalytic Review (1913-1957), 36*(4), 416.

Foucault, M. (1968). Politics and the Study of Discourse. In G. Burchell, C. Gordon, & P. Miller, (Eds.), *(1991) The Foucault Effect, Studies in Governmentality, with Two Lectures by and an Interview with Michel Foucault.* University of Chicago Press.

Foucault, M. (2012). *Discipline and Punish: The Birth of the Prison.* Knopf Doubleday Publishing Group.

Frosh, S. (2001). On reason, discourse, and fantasy. *American Imago, 58*(3), 627–647. https://doi.org/10.1353/aim.2001.0013

Frosh, S. (2013). Psychoanalysis, colonialism, racism. *Journal of Theoretical and Philosophical Psychology, 33*(3), 141–154. https://doi.org/10.1037/a0033398

Gardner, W. L., Karam, E. P., Alvesson, M., & Einola, K. (2021). Authentic leadership theory: The case for and against. *The Leadership Quarterly, 32*(6), 101495. https://doi.org/10.1016/j.leaqua.2021.101495

Glynos, J. (2010). *?Lacan at Work?* In C. Cederstrom & C. Hoedemaeker, (Eds.), Lacan and Organization (pp. 13–58). MayFly Books. https://repository.essex.ac.uk/4085/

Glynos, J. (2011). On the ideological and political significance of fantasy in the organization of work. *Psychoanalysis, Culture & Society, 16*(4), 373–393. https://doi.org/10.1057/pcs.2010.34

Gunder, M. (2005). Obscuring difference through shaping debate: A Lacanian view of planning for diversity. *International Planning Studies, 10*(2), 83–103. https://doi.org/10.1080/13563470500258774

Gunder, M. (2006). Sustainability: Planning's saving grace or road to perdition? *Journal of Planning Education and Research, 26*(2), 208–221. https://doi.org/10.1177/0739456X06289359

Healey, P. (2004). Creativity and urban governance. *disP - The Planning Review, 40*(158), 11–20. https://doi.org/10.1080/02513625.2004.10556888

Hillier, J. (2000). Going round the back? Complex networks and informal action in local planning processes. *Environment and Planning A, 32*, 33–54.

Hillier, J., & Gunder, M. (2003). Planning fantasies? An exploration of a potential lacanian framework for understanding development assessment planning. *Planning Theory, 2*(3), 225–248. https://doi.org/10.1177/147309520323005

Hook, D. (2017). *Six Moments in Lacan: Communication and Identification in Psychology and Psychoanalysis*. Routledge.

Hook, D., & Vanheule, S. (2016). Revisiting the master-signifier, or, Mandela and repression. *Frontiers in Psychology, 6*. www.frontiersin.org/articles/10.3389/fpsyg.2015.02028

Howlett, M., & Ramesh, M. (2014). The two orders of governance failure: Design mismatches and policy capacity issues in modern governance. *Policy and Society, 33*(4), 317–327. https://doi.org/10.1016/j.polsoc.2014.10.002

Jessop, B. (2002). Governance and Meta-governance in the Face of Complexity: On the Roles of Requisite Variety, Reflexive Observation, and Romantic Irony in Participatory Governance. In H. Heinelt, P. Getimis, G. Kafkalas, R. Smith, & E. Swyngedouw (Eds.), *Participatory Governance in Multi-Level Context: Concepts and Experience* (pp. 33–58). VS Verlag für Sozialwissenschaften. https://doi.org/10.1007/978-3-663-11005-7_2

Kaplan, S., & Orlikowski, W. J. (2013). Temporal work in strategy making. *Organization Science, 24*(4), 965–995. https://doi.org/10.1287/orsc.1120.0792

Kapoor, I. (2005). Participatory development, complicity and desire. *Third World Quarterly, 26*(8), 1203–1220. https://doi.org/10.1080/01436590500336849

Kapoor, I. (2015). What 'drives' capitalist development? *Human Geography, 8*(3), 66–78.

Kooij, H.-J., Van Assche, K., & Lagendijk, A. (2014). Open concepts as crystallization points and enablers of discursive configurations: The case of the innovation campus in the Netherlands. *European Planning Studies, 22*(1), 84–100. https://doi.org/10.1080/09654313.2012.731039

Ledeneva, A. (2006). *How Russia Really Works: The Informal Practices That Shaped Post-Soviet Politics and Business*. Cornell University Press.

Ledeneva, A. (2018). *The Global Encyclopaedia of Informality, Volume 1: Towards Understanding of Social and Cultural Complexity*. UCL Press.

Luhmann, N. (1989). *Ecological Communication*. University of Chicago Press.

Luhmann, N. (1990). *Essays on Self-Reference*. Columbia University Press.

Luhmann, N. (1995). *Social Systems* (Vol. 1). Stanford University Press.

Luhmann, N. (2018). *Organization and Decision*. Cambridge University Press.

McGowan, T. (2017). *Capitalism and Desire: The Psychic Cost of Free Markets*. Columbia University Press. https://doi.org/10.7312/mcgo17872

McGowan, T. (2019). Law and Superego. In Y. Stavrakakis (Ed.), *Routledge Handbook of Psychoanalytic Political Theory* (pp. 139–150). Routledge.

Palestrino, M. (2022). Neglected times: Laclau, affect, and temporality. *Journal of Political Ideologies, 27*(2), 226–245. https://doi.org/10.1080/13569317.2021.1916201

Panizza, F., & Stavrakakis, Y. (2020). Populism, Hegemony, and the Political Construction of "The People": A Discursive Approach. In P. Ostiguy, F. Panizza, & B. Moffitt (Eds.), *Populism in Global Perspective*. Routledge.

Penz, O., & Sauer, B. (2019). *Governing Affects: Neoliberalism, Neo-Bureaucracies, and Service Work*. Routledge.

Pohl, L., & Swyngedouw, E. (2023). Enjoying climate change: *Jouissance* as a political factor. *Political Geography, 101*, 102820. https://doi.org/10.1016/j.polgeo.2022.102820

Pusca, A. (2007). Shock, therapy, and postcommunist transitions. *Alternatives, 32*(3), 341–360. https://doi.org/10.1177/030437540703200304

Rasche, A. (Ed.). (2008). Strategic Realities—The Role of Paradox. In *The Paradoxical Foundation of Strategic Management* (pp. 179–191). Physica-Verlag HD. https://doi.org/10.1007/978-3-7908-1976-2_5

Ruti, M. (2012). *The Singularity of Being: Lacan and the Immortal Within*. Fordham University Press.

Scott, J. C. (1998). *Seeing Like a State: How Certain Schemes to Improve the Human Condition Have Failed*. Yale University Press.

Sehring, J. (2009). Path dependencies and institutional bricolage in post-Soviet water governance. *Water Alternatives, 2*(1). 61–81.

Skeggs, B. (2013). *Class, Self, Culture*. Routledge. https://doi.org/10.4324/978131 5016177

Snyder, J. (2015). The Biopolitics of Neoliberal Governance. In J. Snyder (Ed.), *Poetics of Opposition in Contemporary Spain: Politics and the Work of Urban Culture* (pp. 125–162). Palgrave Macmillan US. https://doi.org/10.1057/9781137533210_4

Soler, C. (2015). *Lacanian Affects: The Function of Affect in Lacan's Work* (1st ed.). Routledge. https://doi.org/10.4324/9781315731797

Stapleton, K., & Wilson, J. (2017). Telling the story: Meaning making in a community narrative. *Journal of Pragmatics*, *108*, 60–80. https://doi.org/10.1016/j.pra gma.2016.11.003

Stavrakakis, Y. (2002). *Lacan and the Political*. Routledge.

Swyngedouw, E. (2022). The unbearable lightness of climate populism. *Environmental Politics*, *31*(5), 904–925. https://doi.org/10.1080/09644016.2022.2090636

Teisman, G., van Buuren, A., & Gerrits, L. M. (2009). *Managing Complex Governance Systems*. Routledge.

Vaara, E., Sonenshein, S., & Boje, D. (2016). Narratives as sources of stability and change in organizations: Approaches and directions for future research. *Academy of Management Annals*, *10*(1), 495–560. https://doi.org/10.5465/19416 520.2016.1120963

Valentinov, V., Roth, S., & Will, M. G. (2018). Stakeholder theory: A Luhmannian perspective. *Administration & Society*, *51*(5), 826–849. https://doi.org/10.1177/00953 99718789076

Van Assche, K., Beunen, R., & Duineveld, M. (2013). *Evolutionary Governance Theory: An Introduction*. Springer.

Van Assche, K., Beunen, R., & Gruezmacher, M. (2024). *Strategy for Sustainability Transitions: Governance, Community and Environment*. Edward Elgar Publishing.

Van Assche, K., Devlieger, P., Teampau, P., & Verschraegen, G. (2009). Forgetting and remembering in the margins: Constructing past and future in the Romanian Danube Delta. *Memory Studies*, *2*(2), 211–234. https://doi.org/10.1177/17506 98008102053

Van Assche, K., Gruezmacher, M., & Deacon, L. (2020). Land use tools for tempering boom and bust: Strategy and capacity building in governance. *Land Use Policy*, *93*, 103994–103994. https://doi.org/10.1016/j.landusepol.2019.05.013

Van Assche, K., Gruezmacher, M., Marais, L., & Perez-Sindin, X. (2023). *Resource Communities: Past Legacies and Future Pathways*. Taylor & Francis. https://books. google.ca/books?id=4EHUEAAAQBAJ

Vanheule, S. (2016). Capitalist discourse, subjectivity and Lacanian psychoanalysis. *Frontiers in Psychology*, *7*. https://doi.org/10.3389/fpsyg.2016.01948

Voß, J.-P., & Bornemann, B. (2011). The politics of reflexive governance: Challenges for designing adaptive management and transition management. *Ecology and Society*, *16*(2), 9. [online] URL: www.ecologyandsociety.org-9

Wagenaar, H. (2004). "Knowing" the rules: Administrative work as practice. *Public Administration Review*, *64*(6), 643–656. https://doi.org/10.1111/ j.1540-6210.2004.00412.x

Wagenaar, H. (2015). 22 Transforming perspectives: The critical functions of interpretive policy analysis. In F. Fischer, D. Torgerson, A. Durnová, & M. Orsini, *Handbook of Critical Policy Studies* (pp. 422–440). Edward Elgar Publishing Limited.

Wardle, B. (2016). You complete me: The Lacanian subject and three forms of ideological fantasy. *Journal of Political Ideologies, 21*(3), 302–319. https://doi.org/
10.1080/13569317.2016.1208246

Yanow, D. (2000). *Conducting Interpretive Policy Analysis* (Vol. 47). Sage.

Ziai, A. (2004). The ambivalence of post-development: Between reactionary populism
and radical democracy. *Third World Quarterly, 25*(6), 1045–1060.

Žižek, S. (1991). Formal democracy and its discontents. *American Imago, 48*(2),
181–198.

Žižek, S. (1992). *Looking Awry: An Introduction to Jacques Lacan through Popular
Culture*. MIT Press.

Žižek, S. (2002). *Did Somebody Say Totalitarianism?: Five Interventions in the (mis)
use of a Notion*. Verso.

Žižek, S. (2009). *In Defense of Lost Causes*. Verso. http://public.ebookcentral.proqu
est.com/choice/publicfullrecord.aspx?p=5176960

Žižek, S. (2019). *The Sublime Object of Ideology*. Verso Books.

3 Governance routines and exceptional situations

Coordination and realities, organizing and thinking

Governance cannot work on an ad hoc basis. For citizens, governance makes life more predictable, and this means that rules and policies are required, as well as stories about the world that are not too complicated, and not too strange. For governance, coordinating collective decision-making, and through decision-making, coordinating both individual and collective action, is a necessity with rules (we speak of institutions) active at each level and in each step (Keating et al., 2023; Van Assche et al., 2013). Once rules are in place, once a form of coordination is enabled through them, changing those rules at will is out of the question, because the perception of legitimacy is essential to keep the social order intact, and because predictability and stability make interactions easier (McSwite, 1997). Once things are organized around a particular set of rules, and once rules to change the rules are introduced, one needs to tread carefully when modifying the institutions.

Stable institutions enable predictable processes in politics and administration, and one can speak here of procedures and routines. To address a particular problem, a procedure is at hand, prestructuring the process of investigation, decision-making, implementation. Even where no new questions are formulated, procedures operate, and simple or regularly applied procedures can be called routines. The term "routine" can apply to complex yearly reviews of a policy domain and to lengthy environmental assessments, but also to the organization of office space, or, in the old days, the structuring of time by the appearance of the coffee lady (Van Assche et al., 2024; Wagenaar, 2004).

Routines help to guide collective thinking, to coordinate action, to translate ideas and decisions into action. They also help non-thinking to survive for a long time, to mask problematic thinking, hence, the history of critiques of bureaucracy focusing on the injustice and stupidity caused by impersonal approaches to real-life issues, and the blindness and numbness instilled by repetitive action (Olsen, 2006; Pierre & Peters, 2000). Positively, routines can help blunt reflexivity and sharpness of observation for good reasons. The reason can be efficiency but also the maintenance of productive fictions. Supporting productive fictions can entail supporting a positive transformation

DOI: 10.4324/9781032696676-3

of society (Žižek, 2009). We might not be sure what sustainability is, but pretending we are sure can actually make a difference. We notice in this example the combination of Master Signifiers and routine, together propelling action, based on a shared value, but doing this unhindered by too much reflection – which would expose that nobody knows what a sustainable community would be (Gunder & Hillier, 2009). Routines can thus amplify the power of societies to transform themselves into a desirable direction. And they can do the opposite, if routines remain unquestioned or even unobserved for a long time. Blind spots that are hard to mend, can stem from routines, while routines can systematically underestimate or diminish the power of governance (du Gay & Lopdrup Hjorth, 2024).

Naturalization, the creation of realities that evade scrutiny of their effects and origins, tends to be accompanied and supported by routines (Duineveld et al., 2013). The contingent character of realities thus moves out of sight without any intention, without the intervention of any discourse. Historical and identity narratives function as natural mediators, as they are already involved in the co-creation of reality. If they are already present in, and possibly structuring the governance process, their effect on naturalization is more likely, their interplay with routines more impactful.

Complexity and limits

Complexity reduction enables systems to function. Individuals, organizations, and governance systems cannot operate without it, and routines play their role in it. Enabling coordination is only one of the functions of complexity reduction. We already mentioned the tamping down of excessive reflexivity, and, with that, of observation (Teisman et al., 2009). Governance systems need continuously to think of their own cost, that is, the cost to the community, but also the cost of internal complexity for its smooth functioning (Luhmann, 1990). Routines can combine clarity in thinking with clarity in organizing, with, in each case, clarity and simplicity reducing the cost of governance operations. It remains possible to pile up routines, make them complicated, so complexity can always strike back, or become problematic at the aggregate level (Valentinov, 2014).

Routines work because they coordinate thinking and organizing, through simplification on each side (Czarniawska, 2013). A procedure in administration is by definition a structuring of time, in which each phase, each step is associated with certain activities. Those activities need to be carried out by different individuals or organizations that become responsible. That responsibility can entail the insertion or application of a bit of specialized expertise, which can be associated with an organization, possibly an administrative unit. Such couplings between thinking and organizing are not the only ones, as procedures can also instill a perspective, a mode of recognizing and analyzing problems, of finding solutions (Newman, 2005). Ideas, and their embedding narratives can thus be directly present in the procedures, or they can be indirectly present, by leaving space for their application.

The organization of governance as a process of crystallizing roles, which can be coupled to positions in various procedures, and of delineating the function of different forms of knowledge, is stabilized by the process of creating organizations, in administration, in the legal and economic systems, in civil society (Olsen, 2008; Seim & Søreide, 2009). Once organizations exist, they can function as actors, and play a role in decision-making processes, in a manner that stabilizes those processes more effectively (Van Assche et al., 2013). Specialization of roles and specialization of knowledge support each other in a similar way. Once a role exists, associated with a form of knowledge, the role is a reason for the continuous production of, and a source of legitimacy for that knowledge (Bierschenk & Olivier de Sardan, 2019).

Procedures thus coordinate thinking and organizing, actors and institutions, time and space, by simplifying all of them, legitimizing all of them (Seidl, 2016). The existence of complex governance systems, with myriad actors and institutions, with specified roles for various sorts of knowledge, keeps procedures in place. The more coordinative work procedures undertake, the harder they are to change, without causing disruption (Cornell, 2014). Routines, thus, are to be understood as institutions themselves, as they coordinate actors and institutions, and can have the character of law, policy, or plan. They shape particular forms of questioning and tend to become unquestioned themselves (Clegg et al., 2016). Organizations with a role as actors in governance can start to identify with procedures, with a particular form of interrogating reality, with a narrowly defined domain of reality, and a very specific angle. More complex forms of governance thus import both flexibility and rigidity; as a structured whole, complex routines allow them to solve complex problems, but their internal complexity makes for a proliferation of blind spots and rigidities (de Roo et al., 2012) (see also next chapter).

A slow erosion of critical capacity, uncritical acceptance of roles, rules, and associated forms of knowledge, can be associated with routines left unquestioned. Options for critique become rarer, within the system, and that critique tends to lose its teeth (Smith & Stirling, 2010). In the community, resistance can grow, and even anger, but the complexity of the system makes it hard to know what to target, and the anger can easily be explained away by insiders claiming that the critics simply do not understand.

Mythologies of leadership, of environments with customers easily seduced by something, assumptions on the nature of fashions (and nonrecognition of fashions as fashion), delusions regarding brand value and the specificity of product and producing organization can all contribute to a simplification, even halting, of thinking and observation (Alvesson & Spicer, 2010). This can be fine when action is needed fast, and when fast internal coordination does not allow too many questions. It is helpful when there is a real window of opportunity, and an environment is amenable to what an organization has to offer. It is less interesting when circumstances change, when internal coordination is degraded in the long term, when the capacity to think shrivels, when the leader stumbles and leaves nothing to identify with (Bolden et al., 2023).

Non-thinking and non-observing thus come with benefits and risks. Routines tend to reduce and streamline thinking and organizing, keeping governance away from possibly dangerous limits, while creating different limits through their silent reproduction. Neither the preexisting limits nor the ones induced by routines, are easily observed (Luhmann, 2018). Not recognizing limitations can come with pros. Harnessing the mobilizing power of not-quite-true and not-yet-true stories might become possible through limiting reflexivity (Muhr & Kirkegaard, 2013). The not-too-realistic analyses and unrealistic expectations associated with a policy might have led to positive collective action, bringing the stories closer to the truth (Beunen et al., 2013; Beunen & Van Assche, 2013). Not truly knowing what sustainability or energy transition might look like does not preclude communities from articulating and following policies that make a big difference. The same situation observed through more bleak narratives can inspire desperation and nonaction; limitations can show up in a harsher light, and what might have been transcended looks insurmountable (Crow & Jones, 2018). Procedures and routines can keep productive fictions in place, not just to stabilize governance but also to maintain a functional opacity, a deliberate glossing over of limitations of governance, in manners that can be performative (Žižek, 2008, 2009, 2020).

Observation, self-observation, and self-critique have no natural optimum in governance. One can say with some confidence, however, that maximization is not optimization, that more is not always better (Seidl & Becker, 2006). Procedures and other routines make reproduction easier for governance systems, by selectively instilling non-thinking and non-observation, by assigning a limited space for self-critique. Smooth functioning is a goal in itself, and disappearing into the background contributes to the processes of naturalization and legitimization that can be expected from governance (Foucault, 2012; Snyder, 2015). Continuous self-doubt brings systems closer to dysfunctional states, and this is the case because of interruptions in organizing but also because of doubt thrown onto the system itself (Liff, 2014; McSwite, 2004). Routines can quell such doubt, and they come with paradoxical benefits for the ambition levels of governance, as the opacity introduced produces both unnecessary limits and the power to overcome seemingly hard boundaries.

Dealing with limitations

Managing limitations through procedure can happen in several ways. We mentioned the enshrining of productive fictions through routines and procedures, rendering them relatively immune for self-doubt. A second productive effect of routines can be the systematic ignoring of limits, or the normalization of an insouciance regarding some of them, with failures not routinely observed or condemned, with results interpreted generously as ongoing experiment. This can in effect mean a gradual chipping away at limits, or a stubborn or creative wayfinding around a problem.

More conscious methods include a shifting of routines or procedures governed by meta-rules or meta-procedures, and the acceptance and institutionalization of episodic shift between routine governance and periods of more intense reflection (Rasche & Seidl, 2020; Sørensen, 2006). The reflection can entail a sharpening of observation, a rethinking of ideals, of governance modes, of forms of representation and participation. Meta-rules include rules to deal with exceptions: rules to break the rules, routines, procedures, and allow for more reflection and judgment, or rules that delineate a particular set of recognizable and anticipated exceptions, then subjected to clearly outlined special procedures (Ledeneva, 2018).

Exceptional phases in governance can be more than phases of intensified reflection. Crisis modes are not conducive to much reflection. Meta-rules can guide the switching of phases by anticipation, by legitimizing a set of crisis or reflexivity modes in advance, as well as acceptable circumstances for triggering them (Boin, 2009). One can think of delineation roles, arenas, committees, rules, expanded mandates, which only appear under exceptional circumstances, delineating very carefully, prescribing exactly what counts as an exception and when it ends, what can happen, what is allowed to change (Boin et al., 2008). When shifts are not encoded in institutions, when they simply happen, forced by external circumstances or radical regime changes, they tend to undermine the possibility for long-term perspectives, and for the community to guide itself in a desirable direction.

Several types of situations might force a questioning or partial suspension of routines. We speak of *shocks, conflicts, gradual loss of trust, amplified uncertainty, and grand challenges.* When routines break up, when they do not function anymore or stop producing desirable results, or the usual but now unsatisfactory results, the moment might have come to rethink the construction of realities in governance. Routines and realities can slowly erode, or suddenly collapse. Coordination in the traditional fashion might have become practically impossible, or unable to motivate actors. The persuasive character of policy narrative, of the structuring fantasies, might suddenly evaporate.

Shock and conflict

Shocks are situations that were not predicted by the governance system, leaving it unable to respond. Explanations might be fabricated, ad hoc institutions might appear, yet the legitimacy of the system is harmed. Quick but scattered self-reflection in governance might be triggered, bringing renewed scrutiny of governance in the community. The limitations of the system might be visible for the community first, then in governance, or the other way around (Van

Assche et al., 2022a). Routines that were not questioned, or not observed, can now appear in the spotlight. Assumptions can invite questioning. A variety of problems makes a recovery more complicated. Chaotic regime change aggravates the effects of shock; loss of legitimacy triggers calls for regime change (Pusca, 2007). Experts needed for recovery might have lost all trust.

Shocks can accompany or instigate uprisings when voices are not heard, or are excluded for a long time; they can stem from the ecological environment. Limitations in the understanding of couplings between governance, community, and environment maintain the possibility of an ecological shock (Luhmann, 1989). Material environments are never fully mapped and understood, so either events or their effects will continue to produce surprises. Shocks do not announce themselves; they leave governance speechless and can cause damage that hampers recovery of governance capacity afterwards (Boin et al., 2020; Duineveld et al., 2017). Productive capacity might be destroyed, and expertise, lives, infrastructures, ecosystems, memories, and identities might be restructured in such a way that coordination afterward becomes difficult. The devastations of war are a prime example (Van Assche & Gruezmacher, 2023).

Conflict can be described as a different situation that exposes the limits of governance, the associated narratives and identities. Shocks can cause conflicts and vice versa, as in war, yet they must be distinguished. Shocks are events affecting the governance system; conflicts are processes that stem from disagreements between people, with a great potential to restructure social identities (Van Assche et al., 2022b). The reasons for and the legacies of conflict are usually numerous, and not always conscious. From a psychoanalytic perspective, it is not strange that conflict can spread through discursive worlds. It can colonize other topics, other groups, deepen divisions, and create divisions where they do not exist. The world becomes reorganized and reinterpreted along the lines of the parties, battles, topics, and disagreements structuring the conflicts. What could be previously balanced in governance, as in diverging futures, memories, desires, what could coexist, can do so no more. Governance always requires balancing, avoiding conflict, and includes procedures to manage conflict. Yet, in post-shock conflicts, damage has been inflicted on those system functions. Shared identities and memories cannot be invoked, and common ground might not exist anymore (Feldman, 1991; North et al., 2009).

Interpretations of self and others shift in protracted conflict; the projections and identifications that might have been contributing to stable governance before, are not functional anymore. Conflict can cause a shock of recognition of previously disavowed narratives, a reverting to old stereotypes of self and others, a rewriting of community identity, of identities becoming more conscious of and consciously rejecting previous similarities and identifications (Glynos, 2011; Hinchman & Hinchman, 1997). An awakening of collective trauma can follow (see the next chapters). This, in turn, can block introspection, can limit the careful reexamination of the past that might have led to the conflict.

A way of evading the balancing of conundrums and issues of integration that is only available to communities and their governance systems, not to individuals, is the *non-construction of identity*. The collective Ego, as we have analyzed it, does not always show up, and Ego serves to integrate contrasting desires, ideals, understandings of self and environment, past, present, and possible futures. An integrated Ego allows an individual to orient herself in the world, to strive for a particular quality of life; governance can sustain a collective Ego and allow it to orient itself in the world. If that Ego disintegrates because of conflict or other reasons, one can resort to a simplification of governance. One can revert to simplified narratives of community and governance that can abstain from articulating a community identity, or a narrative about past and future. It is possible *not* to articulate the collective good, as only individual pursuits are counted as real and desirable, and as the attempt to discern collective goods or a collective itself can be felt as an unfair imposition (Lambert, 2019; Swyngedouw, 2022). Or, one can find pride in the greatness, the freedom, the military might, or simply the size of the country, without being willing and able to articulate a direction. One can distinguish between scales and leave the idea of identity to the nation-state, while understanding the other levels as simple administrative operations.

This reduction of tension is a flight out of a collective identity, or a turn to a simple and phantasmatic identity, a version that deprives itself of further development. There are of course less dramatic versions of managing tensions and conflict. The turn away from the possibility to define shared futures and strategize for them (see also the last chapters of this book) creates its own obstacles for a comeback, as meta-narratives are installed that focus on entirely different matters, such as further simplification of governance under the name of efficiency (Hood & Peters, 2004).

The existence of tensions, and potential conflict in governance, is understandable. Not only does governance embody a synthesis of different voices, but it is also a way to compare, confront, and assess those voices and perspectives on reality (Kornberger et al., 2006; Savski, 2020). Sometimes these are different viewpoints that can considered to be rational, if tied to an interest or an assumption; sometimes they are connected to larger narratives or ideologies, so questioning the smaller points is questioning the big ones (Žižek, 2020). A proposed decision might cast a negative light on an identity in terms of race, gender, ethnicity, or as a religion. If a policy is seen in the light of a version of a religious belief supporting it, critique will be dealt with differently. In the other direction, if anything coming from a faction in council or parliament is instinctively rejected since it is coming from "the bad guys," then not much democratic deliberation is possible, as in weighing of arguments and interests, or as in the possibility to come to a new shared narrative.

Conflict thus exposes limits of governance, and it creates new limits. New ideas can emerge in conflict, even new identities, so it needs to be understood as potentially productive as well. Stamping out, eradicating conflict is not possible; managing it is not only possible but strictly necessary. A variety of tools gradually appeared in governance, devised to ignore, manage, transform, and reinterpret conflict. Conflict can be downplayed, resolved in backrooms, recognized only internally, or ignored thanks to performances of success, of unity. They can be outsourced to the community by means of participation exercises where tensions become visible in the community rather than in governance, and in this way responsibility for decisions can be shifted. Or, experts can be made responsible for divisive decisions, with governance circles washing their hands of this. Calls for increased participation often reflect a breakdown of routines or a loss of trust in them.

Trust

Trust can be lost without conflict or shock. Loss of trust can lead to shock and conflict; it can be caused by them, but is not necessarily tied to them. Trust in actors can evaporate, as can trust in the system of governance, its fairness, its efficiency, and its reality and the realities it produces. Routines designed to black-box complex and shifting realities might become mistrusted themselves (Latour, 2004; Luhmann, 2018). Deep belief in procedural realities can blind the governance system for a growing distrust in society. Limits of what can be achieved through governance look different if the whole system is not trusted. If distrust translates into unwillingness to participate, unworkable situations can quickly appear, and more ambitious forms of governance might fall apart fast (Swain & Tait, 2007; Vries, 2014). If people do not believe that the system works, that it is doing what it is supposed to do, that the motives of actors are what they say they are, then they are not willing to follow rules, and they are less than interested in participating in processes of planning or policymaking.

Trust can be lost at once if something unexpected and terrible happens, or it can be lost gradually. The reasons for this can be a decline in the quality of governance, or an impoverishment of enjoyment, a lower quality of life ascribed to poor governance. Scandals can be catalysts, revelations casting a different light on practices and motifs. Corruption can feel like betrayal, not because public money has been squandered but because collective ideals and identities have been betrayed (Pusca, 2007; Rose-Ackerman & Palifka, 2016). Scandals can cause a collapse of the collective Ego; they can signal that it never existed, and that, therefore, the individual Ego must be reexamined.

Narratives are always at play, even if latently, and so are affects (see Chapter 5). Loss of trust can shift narratives, but very common in governance is the loss of trust by unobserved changes in Master Signifiers and narratives. Those can be connected to social identities but not necessarily so. New social identities might be forming, outside the purview of the governance systems, and old narratives can be reinterpreted or rejected without the system noticing it and responding to it (Brans & Rossbach, 1997; Grothe-Hammer, 2022). A classic example is the notion of "democracy," which is itself the kernel of a range of productive fictions, as "the people" do not exist unambiguously and "rule by the people" is not possible in a literal sense and cannot be reduced to a formula. Various models of democracy are possible, and even more varied practices. Thus, evolving notions of democracy can lead to a gradual nonrecognition of the democratic character of the current system.

As always, trust in something external relates to the interpretation of self (Freud, 2014a, 2014b; Lasch, 1992). If citizens rethink what it means to be a citizen, that is, an upstanding (or, just acceptable) member of a particular community. If an individual citizen has ambiguous feelings about membership, then in relating to its grounding mythologies, trust in governance systems representing and reproducing those mythologies will be affected. If there is Voice in governance, that Voice needs to be trusted, in a double sense as noted before: Is this indeed the voice of the community and do I believe in the story? Is it sincere, attractive, realistic? If ambivalence dominates identification with the community, if perhaps after a phase of overidentification, a process of soul-searching ensues, the voice of governance might appear repellent. What repels might be a former version of self (McGowan, 2022; Ruti, 2012).

Building or rebuilding trust is therefore not a technical exercise. Technicalities cannot be taken for granted, cannot be trusted. There are no prescriptions. Norwegian peace-builders in conflict zones cannot rely on recipes, but they can be successful if they understand the context, the nature of what is at stake, and if they carefully delineate spaces for and steps in voicing difference in perspective. Mediators who are trusted themselves can make a difference, that is, if they are a third party that can help to restore trust, first in the possibility of conflict resolution, of sharing power in governance, of governance itself (Vélez-Torres et al., 2022). Third parties can restore trust in the existence of a Big Other, which is not represented by the previous version of governance, but is felt to be present in the ideals so sincerely inspiring the mediators, or in the idealized version of governance in the mediators' home country.

If trust is a problem, routines must shift. Slowing down governance, where necessary by establishing temporary forms of governance focused on conflict resolution, can contribute to the rebuilding of trust. Without minimal levels of trust, governance cannot function, Voice cannot emerge, and the real diversity of possible futures cannot be imagined and accessed by the community. Distrust can inspire nonobservation and non-thinking, dismissal, and it can inspire overthinking and paranoid ascriptions of agency and intention to a system that might have neither intention nor agency (Short, 2011). It can steer individual and collective desire in directions far removed from what are actually possibilities, and toward shocks that undermine governance capacity for a long time.

Uncertainty

Governance is always dealing with uncertainty, even in its most simple form of crime prevention and service provision. Environmental change can amplify uncertainty, as can social and economic change. While each system is adapted to its environment in some way, and imbued with a form of adaptive capacity, this capacity is still constrained by the organizational and cognitive limitations of the governance system (Duit & Galaz, 2008). What cannot be observed and understood, cannot be responded to, while any response must reckon with a set of available and legitimate institutions. Thus, if patterns of change are themselves transforming, the interpretive, adaptive, predictive tools of governance will turn out to be insufficient at some point (Teisman et al., 2009).

The experience of shocks (see above) can inspire fear, distrust, and disappointment among citizens and with those in governance. One response can be to downplay the risk, while others can be to hide it, try to present it as an opportunity, to fake responses, to pretend one is in full control of the situation (Clarke, 2020; Gunder, 2015). These are moments of great rhetoric, of leadership calming down frayed nerves, and of a proliferation of defensive mechanisms, with individuals and with the governance system as a whole. If governance has Voice, this Voice can be mobilized for such purposes, but if disunity and distrust seep in, amplified uncertainty can expose the limits of governability so clearly that Voice disintegrates – as in many personality disorders described in the psychoanalytic literature (Fairbairn, 1994).

Disavowal, to reduce anxiety in society, to evade responsibility in governance, to keep master narratives intact, is a common occurrence when increased uncertainty is felt (Stavrakakis, 2002). Public disavowal can be combined with internal recognition, but it can also be the other way around, with decision-makers rhetorically accepting a problem but refusing to work on it. Intellectualizing is an option, with experts, in modernist fashion, overestimating their powers, and governance actors blindly accepting these misperceptions. Rationalizing and wishing away uncertainty by quickly shifting to available explanations or by signaling action and leadership through very visible changes in policy course, are commonly deployed to appease the public, to give the impression something is being done and something can be done. Such conditions are conducive for the appearance of charismatic leadership, when not the governance system but the quality of the leader is posited as guarantee for the truth of analysis and solution (Iordachi, 2004).

Grand challenges

The rhetorical recognition of big and complex issues affecting the whole society can help shift narratives toward collaboration, toward rapid action in long-term perspectives. Speaking of grand challenges can expose some of the more problematic defense mechanisms sketched above. In an ideal world, a world community would arise, recognizing its shared problems and collaborating to solve them. In the real world, a fragmentary consensus can form, a slow recognition of blind spots and problematic Master Signifiers (Van Assche et al., 2021). But responses to rhetoric of grand challenges, of larger-than-life problems threatening our lives, also include resistance, backlash, and counterstrategy (Van Assche et al., 2024). Neither rhetoric nor an actual shift in governance, away from routines, automatically extinguishes the desire to cling to clear and simple identities, for stories and promises of non-change and vanishing complexity. Reality can be equated with simplicity when any recognition of complexity is nonsense and any expertise is part of a government conspiracy to suppress real and legitimate desire (McGowan, 2017).

First, grand challenges, if publicly recognized, and even if partially and hesitantly acted upon, thus expose limits of governance in a double sense: the complexity, scale and time frames that have to be engaged with are beyond current governance regimes and their possibilities to collaborate, and second, the response of people cannot be steered and predicted well enough to avoid backlash and worse (Jessop, 2002). The best ideas and most urgently needed solutions, even if technically feasible, might not be persuasive at all, which is a problem in democratic systems. The always present risk of backlash does represent an often-unacknowledged limit of governance, because it emphasizes the fragility of any long-term solution that questions or requires modification of lifestyles, values, desires, identities, and because good governance entails an anticipation of this fragility (Eisenstein & McGowan, 2012). Such sensitivity can lead to a willing blindness to policy solutions that might be entirely necessary (Smith & Stirling, 2010).

Governance, as we know now, is always trying to do the impossible, and it must gloss over that impossibility. Psychoanalysis shows us that individuals are, seemingly, not much better off, maintaining identity and ambition in a world that continuously offers friction to whatever desire and narrative is projected and tested, while our inner world is far more fragmented and far less transparent than we care to see. Yet, psychoanalysis has also demonstrated the continued creativity of people to reinvent themselves, and the mutual stabilization of the individual and the group. Individuals and communities need each other, and they need productive fictions to function, some of them shared.

This means that a relentless focus on exposing those fictions, partly through revealing and dwelling on limitations of governance systems, has inherent risks (Žižek, 2019). Balancing opposite desires is not always possible, nor is respecting each and every desire, while not all displacements, projections, or condensations of desire in public discourse can be observed and responded to. Yet, governance cannot openly recognize that this is the case. Some master narratives are highly problematic, but exposing this comes at a political risk (Bryan, 2016).

A fiction of unity, of compatibility of desires and narratives might be required to work on the Grand Challenges facing our society today (Eberle, 2019; Foroughi, 2020). In line with Niccolo Machiavelli, one might argue for guarding the difference between internal and external discourse by leadership, for a strategic insincerity, a strategically harder distinction between messaging about problems and internal understanding about those problems. Of course, ethical questions can be raised, and a high price might have to be paid. The belief in democracy might be undermined, even if there is a sincere intention by leadership to give it back to the people in better health. Paradoxically, the understandings of self and environment required to gradually whittle away at the complexity of Grand Challenges cannot be imposed on the community. For strategies tackling grand challenges to work, a measure of public understanding, and at the very least public acceptance must exist, yet public acceptance can come with a heavy price tag, one possibly involving false promises, fake simplicity, and sham participation.

Concluding

Routines and procedures in governance come with great benefits and a variety of problems. They simplify thinking and organizing, enable coordination, naturalize and legitimize; they provide focus and structure, while introducing selective blindness and insensitivity to context, goal, and effect. They enable structural attempts to overcome limitations yet impose others that are less observable. Reduction of complexity is at the heart of their functioning, but complexity has a rather nasty habit of striking back (Jessop, 1997; Roo & Hillier, 2016). Societies and their governance systems therefore have evolved a pallet of tactics and strategies to modify, suspend, and question routines, often including meta-narratives and rules that establish episodes of routine functioning and others. Those others can revolve around increased reflexivity, but also around a temporary reduction, and sharper focus, of thinking and observing.

Communities are faced with types of situations that force a response, a questioning or alteration of routines, as they stop producing the desirable results, and this eventually cannot be denied anymore. Shocks, conflicts, increased uncertainty, loss of trust, and Grand Challenges all expose limits of governance and test it severely. They must be distinguished as separate phenomena, but in practice they entangle in many ways that aggravate the problem of responding to them. Thinking and organizing can be dramatically affected by shocks, which cause conflicts, which undermine governance capacity, increasing uncertainty further and rendering future trust building a more formidable challenge.

Democratic mythologies of unified communities and desires, of increased participation as a panacea, can work under certain conditions, can do their magic as productive fiction, but not as a panacea. Incompatible and untransparent desires, latent Master Signifiers and identifications make for unpredictable public responses to any attempt at imposing complexity and unwanted changes or unimagined futures (Glynos & Stavrakakis, 2008; Laclau, 2005). However, political entrepreneurs tend to be ready to grab an opportunity and grab power, further entrenching narratives that might perpetuate conflict, or jeopardize long-term solutions for a grand challenge. Simple solutions are simply not available, as the issues stem from our own lack.

Yet, both the installation of routines and deviation from them offer broad avenues to manage what must be managed, to preserve and transform community, to maintain Voice and the capacity to imagine and organize for alternative futures. When exceptional episodes in governance are allowed to break many routines in a regulated manner, they can become functional moments of rapid response *or* intensified self-reflection. In the second case, the community, through the input of different voices, through the reinterpretation of memory, through more direct engagement with different environments, can shift the interpretation of self, and then come to new ways to orient itself toward the future. In such a process, routines do not have to be abolished, but can be bracketed for a while and later either reinstalled or modified. We develop these ideas in the chapters on strategy and therapy. If the connection between exception and a new normal is not understood, then the exceptional episode will most likely be destabilizing, rather than functioning as rejuvenation, reorientation, or therapy.

References

Alvesson, M., & Spicer, A. (2010). *Metaphors We Lead By: Understanding Leadership in the Real World*. Routledge.

Beunen, R., & Van Assche, K. (2013). Contested delineations: Planning, law, and the governance of protected areas. *Environment and Planning A, 45*(6), 1285–1301.

Beunen, R., Van Assche, K., & Duineveld, M. (2013). Performing failure in conservation policy: The implementation of European Union directives in the Netherlands. *Land Use Policy, 31*, 280–288. https://doi.org/10.1016/j.landusepol.2012.07.009

Bierschenk, T., & Olivier de Sardan, J.-P. (2019). How to study bureaucracies ethnographically? *Critique of Anthropology, 39*(2), 243–257. https://doi.org/10.1177/0308275X19842918

Boin, A. (2009). 13. Crisis leadership in terra incognita: Why meaning making is not enough. In P. 't Hart & K. Tindall (Eds.), *Framing the Global Economic Downturn* (pp. 309–314).

Boin, A., Ekengren, M., & Rhinard, M. (2020). Hiding in plain sight: Conceptualizing the creeping crisis. *Risk, Hazards & Crisis in Public Policy, 11*(2), 116–138. https://doi.org/10.1002/rhc3.12193

Boin, A., Mcconnell, A., & Hart, P. 't. (2008). Governing after Crisis. In A. McConnell, A. Boin, & P. 't Hart (Eds.), *Governing after Crisis: The Politics of Investigation, Accountability and Learning* (pp. 3–30). Cambridge University Press. https://doi.org/10.1017/CBO9780511756122.001

Bolden, R., Gosling, J., & Hawkins, B. (2023). *Exploring Leadership: Individual, Organizational, and Societal Perspectives*. Oxford University Press.

Brans, M., & Rossbach, S. (1997). The autopoiesis of administrative systems: Niklas Luhmann on public administration and public policy. *Public Administration, 75*(3), 417–439.

Bryan, T. K. (2016). Capacity for climate change planning: Assessing metropolitan responses in the United States. *Journal of Environmental Planning and Management, 59*(4), 573–586.

Clarke, M. (2020). Eyes wide shut: The fantasies and disavowals of education policy. *Journal of Education Policy*, *35*(2), 151–167. https://doi.org/10.1080/02680 939.2018.1544665

Clegg, S., e Cunha, M. P., Munro, I., Rego, A., & de Sousa, M. O. (2016). Kafkaesque power and bureaucracy. *Journal of Political Power*, *9*(2), 157–181. https://doi.org/ 10.1080/2158379X.2016.1191161

Cornell, A. (2014). Why bureaucratic stability matters for the implementation of demo-cratic governance programs. *Governance*, *27*(2), 191–214. https://doi.org/10.1111/ gove.12037

Crow, D., & Jones, M. (2018). Narratives as tools for influencing policy change. *Policy & Politics*, *46*(2), 217–234. https://doi.org/10.1332/030557318X15230061022899

Czarniawska, B. (2013). *A City Reframed: Managing Warsaw in the 1990's*. Routledge. https://doi.org/10.4324/9781315080048

de Roo, G., Hillier, J., & Wezemael, J. van. (2012). *Complexity and Planning Systems, Assemblages and Simulations*. Ashgate.

du Gay, P., & Lopdrup Hjorth, T. (2024). Organizing States: The continuing relevance of formal organization within political administration. *Organization Theory*, *5*(1), 26317877241235944. https://doi.org/10.1177/26317877241235944

Duineveld, M., Van Assche, K., & Beunen, R. (2013). Malta's unintentional consequences: Archaeological heritage and the politics of exclusion in the Netherlands. *Public Archaeology*, *12*(3), 139–154. https://doi.org/10.1179/1465 518714Z.00000000039

Duineveld, M., Van Assche, K., & Beunen, R. (2017). Re-conceptualising polit-ical landscapes after the material turn: A typology of material events. *Landscape Research*, *42*(4), 375–384. https://doi.org/10.1080/01426397.2017.1290791

Duit, A., & Galaz, V. (2008). Governance and complexity—Emerging issues for gov-ernance theory. *Governance*, *21*(3), 311–335.

Eberle, J. (2019). Narrative, desire, ontological security, transgression: Fantasy as a factor in international politics. *Journal of International Relations and Development*, *22*(1), 243–268. https://doi.org/10.1057/s41268-017-0104-2

Eisenstein, P., & McGowan, T. (2012). *Rupture: On the Emergence of the Political*. Northwestern University Press.

Fairbairn, W. R. D. (1994). *Psychoanalytic Studies of the Personality*. Psychology Press.

Feldman, A. (1991). *Formations of Violence: The Narrative of the Body and Political Terror in Northern Ireland*. University of Chicago Press. https://press.uchicago.edu/ ucp/books/book/chicago/F/bo3644948.html

Foroughi, H. (2020). Collective memories as a vehicle of fantasy and identifica-tion: Founding stories retold. *Organization Studies*, *41*(10), 1347–1367. https://doi. org/10.1177/0170840619844286

Foucault, M. (2012). *Discipline and Punish: The Birth of the Prison*. Knopf Doubleday Publishing Group.

Freud, S. (2014a). *On Narcissism: An Introduction*. Read Books Ltd.

Freud, S. (2014b). *Wit And Its Relation To The Unconscious*. Routledge. https://doi. org/10.4324/9781315830759

Glynos, J. (2011). On the ideological and political significance of fantasy in the organ-ization of work. *Psychoanalysis, Culture & Society*, *16*(4), 373–393. https://doi.org/ 10.1057/pcs.2010.34

Glynos, J., & Stavrakakis, Y. (2008). Lacan and political subjectivity: Fantasy and enjoyment in psychoanalysis and political theory. *Subjectivity, 24*(1), 256–274. https://doi.org/10.1057/sub.2008.23

Grothe-Hammer, M. (2022). The Communicative Constitution of the World: A Luhmannian View on Communication, Organizations, and Society. In J. Basque, N. Bencherki, & T. Kuhn (Eds.), *The Routledge Handbook of the Communicative Constitution of Organization*. Taylor & Francis. https://doi.org/10.4324/978100 3224914-7

Gunder, M. (2015). The Role of Fantasy in Public Policy Formation. In R. Beunen, K. Van Assche, & M. Duineveld (Eds.), *Evolutionary Governance Theory: Theory and Applications* (pp. 143–154). Springer International Publishing. https://doi.org/ 10.1007/978-3-319-12274-8_10

Gunder, M., & Hillier, J. (2009). *Planning in Ten Words or Less: A Lacanian Entanglement with Spatial Planning*. Ashgate Publishing, Ltd.

Hinchman, L. P., & Hinchman, S. (1997). *Memory, Identity, Community: The Idea of Narrative in the Human Sciences*. SUNY Press.

Hood, C., & Peters, G. (2004). The middle aging of new public management: Into the age of paradox? *Journal of Public Administration Research and Theory, 14*(3), 267–282.

Iordachi, C. (2004). *Charisma, Politics and Violence: The Legion of the "Archangel Michael" in Inter-War Romania*. Trondheim Studies on East European Cultures and Societies.

Jessop, B. (1997). The Governance of Complexity and the Complexity of Governance: Preliminary Remarks on Some Problems and Limits of Economic Guidance. In A. Amin & J. Hausner (Eds.), *Beyond Market and Hierarchy* (pp. 95–128). Edward Elgar Publishing. https://doi.org/10.4337/9781035303410.00012

Jessop, B. (2002). Governance and Meta-governance in the Face of Complexity: On the Roles of Requisite Variety, Reflexive Observation, and Romantic Irony in Participatory Governance. In H. Heinelt, P. Getimis, G. Kafkalas, R. Smith, & E. Swyngedouw (Eds.), *Participatory Governance in Multi-Level Context: Concepts and Experience* (pp. 33–58). VS Verlag für Sozialwissenschaften. https://doi.org/10.1007/978-3-663-11005-7_2

Keating, M., McAllister, I., Page, E. C., & Peters, B. G. (2023). *The Problem of Governing: Essays for Richard Rose*. Springer Nature.

Kornberger, M., Clegg, S. R., & Carter, C. (2006). Rethinking the polyphonic organization: Managing as discursive practice. *Scandinavian Journal of Management, 22*(1), 3–30. https://doi.org/10.1016/j.scaman.2005.05.004

Laclau, E. (2005). *On Populist Reason*. Verso.

Lambert, A. (2019). Psychotic, acritical and precarious? A Lacanian exploration of the neoliberal consumer subject. *Marketing Theory, 19*(3), 329–346. https://doi.org/ 10.1177/1470593118796704

Lasch, C. (1992). The Culture of Narcissism. In R. Wilkinson (ed.), *American Social Character* (pp. 241–268). Routledge.

Latour, B. (2004). *Politics of Nature: How to Bring the Sciences into Democracy*. Harvard University Press. https://doi.org/10.4159/9780674039964

Ledeneva, A. (2018). *The Global Encyclopaedia of Informality, Volume 1: Towards Understanding of Social and Cultural Complexity*. UCL Press.

Liff, R. (2014). Unintended consequences of NPM drive the "bureaucracy." *International Journal of Public Administration, 37*(8), 474–483. https://doi.org/10.1080/01900 692.2013.865644

Luhmann, N. (1989). *Ecological Communication*. University of Chicago Press.

Luhmann, N. (1990). *Political Theory in the Welfare State*. De Gruyter.

Luhmann, N. (2018). *Organization and Decision*. Cambridge University Press.

McGowan, T. (2017). *Capitalism and Desire: The Psychic Cost of Free Markets*. Columbia University Press. https://doi.org/10.7312/mcgo17872

McGowan, T. (2022). Sublimating the commodity. *Distinktion: Journal of Social Theory, 23*(2–3), 359–377. https://doi.org/10.1080/1600910X.2022.2054447

McSwite, O. C. (1997). Jacques Lacan and the theory of the human subject: How psychoanalysis can help public administration. *American Behavioral Scientist, 41*(1), 43–63. https://doi.org/10.1177/0002764297041001005

McSwite, O. C. (2004). Creating reality through administrative practice: A Psychoanalytic reading of Camilla Stivers' bureau men, settlement women. *Administration & Society, 36*(4), 406–426. https://doi.org/10.1177/0095399704266744

Muhr, S. L., & Kirkegaard, L. (2013). The dream consultant: Productive fantasies at work. *Culture and Organization, 19*(2), 105–123. https://doi.org/10.1080/14759 551.2011.644670

Newman, S. (2005). *Power and Politics in Poststructuralist Thought: New Theories of the Political*. Routledge. https://doi.org/10.4324/9780203015490

North, D. C., Wallis, J. J., & Weingast, B. R. (2009). *Violence and Social Orders: A Conceptual Framework for Interpreting Recorded Human History*. Cambridge University Press.

Olsen, J. P. (2006). Maybe it is time to rediscover bureaucracy. *Journal of Public Administration Research and Theory, 16*(1), 1–24. https://doi.org/10.1093/jopart/mui027

Olsen, J. P. (2008). The ups and downs of bureaucratic organization. *Annual Review of Political Science, 11*, 13–37. https://doi.org/10.1146/annurev.polisci.11.060 106.101806

Pierre, J., & Peters, G. B. (2000). *Governance, Politics and the State*. Edward Elgar.

Pusca, A. (2007). Shock, therapy, and postcommunist transitions. *Alternatives, 32*(3), 341–360. https://doi.org/10.1177/030437540703200304

Rasche, A., & Seidl, D. (2020). A Luhmannian perspective on strategy: Strategy as paradox and meta-communication. *Critical Perspectives on Accounting, 73*, 101984–101984. https://doi.org/10.1016/j.cpa.2017.03.004

Roo, G. de, & Hillier, J. (2016). *Complexity and Planning: Systems, Assemblages and Simulations*. Routledge.

Rose-Ackerman, S., & Palifka, B. J. (2016). *Corruption and Government: Causes, Consequences, and Reform*. Cambridge University Press.

Ruti, M. (2012). *The Singularity of Being: Lacan and the Immortal Within*. Fordham University Press.

Savski, K. (2020). Polyphony and polarization in public discourses: Hegemony and dissent in a Slovene policy debate. *Critical Discourse Studies, 17*(4), 377–393. https://doi.org/10.1080/17405904.2019.1609537

Seidl, D. (2016). *Organisational Identity and Self-Transformation: An Autopoietic Perspective*. Routledge.

Seidl, D., & Becker, K. H. (2006). Organizations as distinction generating and processing systems: Niklas Luhmann's contribution to organization studies. *Organization*, *13*(1), 9–35. https://doi.org/10.1177/1350508406059635

Seim, L. T., & Søreide, T. (2009). Bureaucratic complexity and impacts of corruption in utilities. *Utilities Policy*, *17*(2), 176–184. https://doi.org/10.1016/j.jup.2008.07.007

Short, J. L. (2011). The paranoid style in regulatory reform. *Hastings LJ*, *63*, 633.

Smith, A., & Stirling, A. (2010). The politics of social-ecological resilience and sustainable socio-technical transitions. *Ecology and Society*, *15*(1). www.jstor.org/stable/26268112

Snyder, J. (2015). The Biopolitics of Neoliberal Governance. In J. Snyder (Ed.), *Poetics of Opposition in Contemporary Spain: Politics and the Work of Urban Culture* (pp. 125–162). Palgrave Macmillan US. https://doi.org/10.1057/9781137533210_4

Sørensen, E. (2006). Metagovernance: The changing role of politicians in processes of democratic governance. *The American Review of Public Administration*, *36*(1), 98–114. https://doi.org/10.1177/0275074005282584

Stavrakakis, Y. (2002). *Lacan and the Political*. Routledge.

Swain, C., & Tait, M. (2007). The crisis of trust and planning. *Planning Theory & Practice*, *8*(2), 229–247. https://doi.org/10.1080/14649350701324458

Swyngedouw, E. (2022). The unbearable lightness of climate populism. *Environmental Politics*, *31*(5), 904–925. https://doi.org/10.1080/09644016.2022.2090636

Teisman, G., van Buuren, A., & Gerrits, L. M. (2009). *Managing Complex Governance Systems*. Routledge.

Valentinov, V. (2014). The complexity–sustainability trade-off in Niklas Luhmann's social systems theory. *Systems Research and Behavioral Science*, *31*(1), 14–22.

Van Assche, K., Beunen, R., & Duineveld, M. (2013). *Evolutionary Governance Theory: An Introduction*. Springer.

Van Assche, K., Beunen, R., & Gruezmacher, M. (2024). *Strategy for Sustainability Transitions: Governance, Community and Environment*. Edward Elgar Publishing.

Van Assche, K., & Gruezmacher, M. (2023). Remembering Ypres. Post-war reconstruction, land and the legacies of shock and conflict. *Land*, *12*(1). https://doi.org/10.3390/land12010021

Van Assche, K., Gruezmacher, M., & Beunen, R. (2022a). Shock and conflict in social-ecological systems: Implications for environmental governance. *Sustainability*, *14*(2), 610. https://doi.org/10.3390/su14020610

Van Assche, K., Gruezmacher, M., & Beunen, R. (2022b). Why governance is never perfect: Co-evolution in environmental policy and governance. *Sustainability*, *14*(15), Article 15. https://doi.org/10.3390/su14159441

Van Assche, K., Gruezmacher, M., Vodden, K., Gibson, R., & Deacon, L. (2021). Reinvention paths and reinvention paradox: Strategic change in Western Newfoundland communities. *Futures*, *128*, 102713–102713. https://doi.org/10.1016/j.futures.2021.102713

Vélez-Torres, I., Gough, K., Larrea-Mejía, J., Piccolino, G., & Ruette-Orihuela, K. (2022). "Fests of vests": The politics of participation in neoliberal peacebuilding in Colombia. *Antipode*, *54*(2), 586–607. https://doi.org/10.1111/anti.12785

Vries, J. de. (2014). *Understanding Trust: Longitudinal Studies on Trust Dynamics in Governance Interactions*. Wageningen University.

Wagenaar, H. (2004). "Knowing" the rules: Administrative work as practice. *Public Administration Review*, *64*(6), 643–656. https://doi.org/10.1111/j.1540-6210.2004.00412.x

Žižek, S. (2008). *Violence*. Picador.
Žižek, S. (2009). *In Defense of Lost Causes*. Verso. http://public.ebookcentral.proqu est.com/choice/publicfullrecord.aspx?p=5176960
Žižek, S. (2019). *The Sublime Object of Ideology*. Verso Books.
Žižek, S. (2020). *The Plague of Fantasies*. Verso Books.

4 Reality and reality testing in governance

We are familiar now with the mediating role of governance in reality construction. Not everyone agrees: in modernist approaches, reality is simply there, citizens and their preferences are simply there, and good governance would consist of legitimate responses to these citizens expressing their preferences. We understand governance as potentially expressing much more. Governance systems do not start reality construction from scratch, but they have reasons to build their own understanding of society and environment, and reasons to try to shape the realities as perceived and felt by citizens. Governance systems have incentives to veil the imperfect nature of their functioning and the contingent nature of both their structure and the decisions taken in those structures.

Continuous input is needed to know whether the understanding of reality that governance operates upon makes sense, and does not diverge too much from what citizens experience as real, while retaining spaces for the realities of experts, the possible realities leaders propose, and the realities of governance itself, its necessities and limitations. Governance is thus, as noted before, an a priori impossible task of coordinating realities. One way to manage this impossibility is to construct realities that can connect others, narratives that can synthesize, gloss over differences, and possibly transform identities, which then perceive realities in new and more bridging ways. Uncertainty always remains.

Learning

Governance systems learn, which helps them deal with new situations. The learning possibilities for governance systems are constrained, however, as they are contingently structured through their experiences. Each system is cognizant of the world in a limited manner and can transform itself only in a few ways that are not destabilizing, that do not undermine its coordinative capacity. We categorized those mechanisms and distinguished between comparative learning, expert learning, experimental learning, reflexive learning,

DOI: 10.4324/9781032696676-4

and dialectic learning (Van Assche et al., 2020). Each of these modes starts from an understanding of reality and transforms that understanding.

Comparative learning is learning by looking at others, at other governance systems. Actors in governance ask themselves, what might count as "best practices," what made for success when pursuing a similar policy goal elsewhere or a different goal in our system? Comparative learning can go deeper, when analyzing how elsewhere goals are articulated, how problems are defined and ranked, and which forms of knowledge and modes of investigation are embraced (Van Assche et al., 2020). Deep comparison becomes aware of not only the differences in governance systems, but also the differences in their self-transformation and learning options. A focus on the comparison of solutions and tools is shallower, and often more problematic. A tool might have worked somewhere, a solution might have satisfied, but this might not be telling about its suitability here. Comprehensive comparison is not possible, as a full description of the observed system is not possible, and as the influence of the observing system is always there. Thus, simple comparison is not that useful, full comparison is out of the question, and the selectivity of observation of the other, with the intent of learning, is influenced by selectivities in the observing remembering self (Luhmann, 1995b).

Thus, the history of the system, its memories and legacies, affect what can be observed elsewhere, as well as the intentions and hopes involved (see Chapter 2). What one is willing and able to observe and to learn hinges on the patterns of identification at play, on versions of reality and its problems circulating in the governance system, on memories and wishes for the future (Okulicz-Kozaryn & Valente, 2018). How governance systems learn through comparison thus reveals much of their personality and their relations to other communities, or to the particular one providing focus for the comparison. Admiration, envy, indifference, even fear, can be at work, while the pressure to differentiate and compete with others feeds into communal narcissism (Lasch, 1992). Accepting that better solutions might be found elsewhere, that qualities elsewhere might shed a light on problems at home, is not easy, and pseudo-technical techniques of comparison do not eradicate the issue.

Neoliberal ideologies, but also ideals of communitarianism emphasizing the local, can encourage such narcissism, creating a drive to confirm one's superiority and difference by copying the supposedly best solutions from many different places. The same narcissism can make it very difficult to observe why something actually worked somewhere, and, more fundamentally, what constitutes a quality elsewhere, something worth learning about and learning from (Tingle, 1992).

Reflexive learning is subjected to the same dynamics, as learning from self-observation, from one's own past, always starts from an imperfectly observed reality. History leaves traces, in modes of observing, and it produces an observing self that does so by using stories, ideas, affects, and Master Signifiers ultimately derived from encounters with others. The observation of self and the distinctions with others cannot escape the self. Reflexivity, as the capacity of a governance system to take a step back and look at its own routines, narratives, hierarchies, and histories from a different perspective, as the capacity to learn lessons and modify behavior based on such reinterpretation, is therefore not a simple capacity (Voß & Bornemann, 2011). Reflexivity does not undo trauma, does not banish the unconscious, and cannot ignore the fact that desire comes from somewhere else.

Specialized roles in governance can be assigned to enhance reflexivity of the system. New stories about the self are more likely to emerge, and second-order observation can be institutionalized in this way (Van Assche et al., 2024). Drawing lessons, however, remains difficult. Those involved in or responsible for the normal routines cannot easily let go, cannot be induced readily to follow the story painted by those engaged in second-order observation. Self-reflection is not easy in governance systems operating largely on routines and delicate balancing. Deep reflexive learning entails a willingness to question values, certainties, and identities, notions of practicality and normality. Reflexivity, in other words, is a headache for people and organizations just trying to do their job (Alvesson & Spicer, 2016). The perceived potential of reflexivity to continuously reshuffle routines and roles, to undermine values and identities is not an incentive even to get started, for many involved. For evasive tactics to be avoided, reflexive learning must be shouldered by the whole system, even if specialized roles exist, while no specialized role can entirely escape the limitations of the system.

Expert learning is technically easier, as it does not require second-order observation distributed throughout the system. Experts can be given clear roles, with clear questions being asked, and carefully delineated application of their expert perspective, which can then be assigned to one step within recognized procedures. Procedures can combine and coordinate knowledge, but not necessarily with much discussion or reflexivity. As one might expect, the selection of issues where expertise is accepted, that is, the selection of methods, theories, and concepts allowed in the system, cannot be extricated from the history of governance, its master narratives, its overriding desires and fears (Fischer, 2000; Latour, 2004). What would be recognized as expertise, how it might be valued, and which use for it could be imagined, hinges on narrative understandings of self and environment, of histories of problem definition and problem-solving based on those understandings.

Experts reinforcing the reigning ideology find easier access to governance, and once their expertise is entrenched, in the organization, and in the institutions produced and used in governance, it is harder to get rid of, as it contributes to the self-stabilization of the discursive configuration (Strassheim & Kettunen, 2014). Ideologies naturalized through governance represent a world both believed and desired, and expertise will be harnessed that buttresses this safe world. Major problems facing the community will then be interpreted according to the mutually reinforcing frames of ideology and expertise. It can happen that expertise *is* the ideology, which often entails that it functions as symbol and guarantor of good governance, that too much is expected of it (Carr, 2010; Scott, 2020). Application and accumulation of expertise can substitute for actual thinking, discussing, and problem-solving.

Experimental learning, in contrast, rarely occurs. Emulating laboratory situations, trying something under controlled conditions, and drawing lessons is not a natural phenomenon. In governance, keeping all other factors constant and gauging the influence of only one parameter is virtually impossible (Hood & Peters, 2004). Trying something is not the same as experimenting, as one cannot expect to identify the factors contributing to success or failure. We mention experimental learning, however, because it is currently in fashion, meaning that there is a belief that people will try it and that there will be effects. Medical experts, via economics, promoted the use of policy experiments, while neoliberal discourses on policy learning and innovation spread the belief that experimentation could identify best practices, ready for diffusion to other contexts (Howe, 2004).

Dialectic learning is learning through discussion, the discovery and construction of new ideas through critical engagement. Dialectic learning is more productive if the other forms of learning are present in governance (Van Assche, Gruezmacher et al., 2022). It can be presented as fairly irreplaceable. Democracy thrives on dialectic learning, which has to be carefully distinguished from deliberation (McAfee, 2022). Adding up wish lists, analyzing preferences, objectively balancing interests is not only futile, but also the belief in its possibility renders actual discussion unlikely. Dialectic learning is helpful in finding common ground, in establishing new perspectives on past and future, on opportunities and threats.

Policy problems and demands do not exist as neutral facts, and where dialectic learning is common practice, this is blindingly obvious, while elsewhere, modernist governance narratives firmly instill the myth of objective policy optimization, while silently undermining the belief in dialectic learning (Ferguson, 1994). What looks like a problem, depends on what people value and desire, and what they value is not akin to what they need (Rochefort & Cobb, 1994). Following Lacan, we distinguish between need, demand, and desire. Very rarely do individuals know what they need, and even more rarely do communities know this. What they demand tends to differ from both need and desire (Fink, 1999). Demands are what is formulated, which thus relies on the sign systems available for the formulation, the occasions and rules shaping the pronouncement in governance, and the self-knowledge with regard to desires. Thus, demand tends to differ significantly from what is desired, and what is needed in a material, psychological, or spiritual sense might even be less accessible than desire (see also Chapter 5).

Desire shifts its object and is never satisfied. Communities become dissatisfied when things go well, when they discover that their old desire is now located somewhere else, or when they find new desires, and they get bored when affective investment in once popular stories and warmly embraced leaders dwindles (Ruti, 2008). Desire thus creates instability in governance (see Chapter 5). New ideals, new objects of desire appear, new expectations that are impossible to satisfy (Gunder & Hillier, 2009). Free-floating desire can fuel enthusiasm for bold utopian schemes, but it can also attach to small events, things, and behaviors, which suddenly come to symbolize a new future, a new identity. Conversely, both big visions and small changes can attract disproportionately negative reactions, when they are felt to infringe upon a desire central to the subject, whether that desire is conscious or not, whether it is deemed legitimate or not (Clegg et al., 2017).

Dialectic learning represents the most likely site for a managed transformation of desire, the transformation of desire through communication, through structured yet cognitively open interaction with others (Wagenaar, 2015). Reinterpretation is the key word, as dialectical learning, when it is not felt as mere negotiation, can entail the reinterpretation of cherished narratives, of formerly coolly approached topics, and formerly apprehensively observed identities. Others' affects can be understood and felt, new objects of desire discovered. New narratives and bonds can be forged. None of this is manageable as it is unpredictable, yet the context of governance can provide structure and pressure to engage with others, if the need to come to decisions is clear, if all parties feel they belong to the same community, and if the possibility of synergies is believable.

Persuasion is a close relative of dialectical learning and is usually involved in its work. Creativity often shows up in its activities. Indeed, the dialectic creation of new stories, about the present, about the "real" issues and desires of those considering a problem can connect different desires, or make them more similar (Gunder, 2005; Healey, 2004). Sublimation can find a home in moments of dialectical learning, where initial obstacles for collaboration are overcome, where initial motivations are transformed, are endowed with creative potential (Kim et al., 2013). Visions for the future forged through dialectic learning can, sometimes, inspire in this very particular sublimating sense.

Alas, this is not always the case, and all too often, motivations remain the same and remain focused on private gain. Learning in governance includes myriad forms of *dark learning*. We understand as dark learning those forms of learning that serve no collective good but only individual and factional goals. This includes the acquisition of knowledge about the system, its routines, its modes of observation, its hierarchies, its weak spots for corruption, for stupidity, the difference between its actual limits and the self-understanding of those limits. Dark learning encourages non-identification with the governance system, with the officially proclaimed values and public goods, while non-identification opens the door more easily for dark learning. Paradoxically, an amoral stance vis-à-vis the governance system, which is useful for dark learning, is also helpful in recognizing the fictions circulating in the system. What we have declared so useful to enhance reflexivity and adaptive capacity, and so hard to achieve, comes more easily to those who do not care.

Blind spots

Learning has its limits, and those limits are diverse. Non-learning stems not only from what cannot be observed by the system, but also from willful non-observation and non-interpretation. Non-learning can be a result of an unconscious resistance to learn, even a traumatic impossibility to remember or observe. Those circumstances deserve understanding, maybe therapy, rather than condemnation (see Chapters 5, 11, and 12). Non-observation, non-thinking, and non-learning also serve positive functions, just as forgetting is constitutive for remembering (Esposito, 2008). Non-thinking can guide and sharpen thinking, as thinking can only work, can only lead to insight and guide observation and learning, if there are structures and limits. As with any boundary, the structuring can emanate from the outside or the inside – an intentional focus, an unseen limit, an unobserved contingent structure. Non-thinking, furthermore, connects thinking to social identities, which have to obscure their own contingency and require a series of blind spots to function (Vanheule, 2016).

Each discourse produces blind spots, as not everything can be observed or emphasized, as not every distinction can be made to structure other distinctions. Blind spots can be intentional at first, then forgotten, and they can be the effect of observational limits that might find their origin in distinctions far removed from the blind spots (Foucault, 2003; Luhmann, 1995a). Stability is not a feature of blind spots, neither for people, nor for organizations, and certainly not for governance systems. They move around and become less accessible for insiders and outsiders alike in that process. Sometimes, however, deeply entrenched blind spots can open up, by topics in governance appearing as relevant, by a reexamination of policy tools, by a period of self-questioning caused by cultural or ideological change (Johnston, 2005).

Shifting discourse and new hybridizations of knowledge move blind spots around, new events occur which can leave blind spots intact or which are so much at odds with current interpretations that they expose or shift blind spots (Bal, 2002; Eco, 1979). Creatively, new events, in the discursive or material worlds, can create new blind spots, where new explanations have to be found, which nevertheless could not be allowed to reveal what was truly unsayable, or simply unobservable. What is unobserved thus can be the result of a limit of observation and cognition, from a focus and structure in thinking and seeing, and from deliberate non-seeing and thinking (Ruti, 2012; Žižek, 2009). Avoiding, evading, and hiding are accompanied by, and possible because of, blind spots. In not confronting an issue, in not questioning non-observation when there is pressure to look more directly, the response, inspired by the desire not to see, can be a move to another discourse, riddled with blind spots, which leave the first one largely intact. New stories can explain away old injustices in new ways, without really addressing them, or the guilt associated with them (Frosh, 2013).

Blind spots, therefore, are not only limits of observation stemming from cognitive limits, from strategies and compromises in governance. They can stem from virtually any operation in governance, can serve several functions. As governance systems, organizations, and individuals all avail themselves of discourse and cannot distinguish themselves from discourse, and as governance systems can develop their own Voice yet remain entangled with organizations and individuals, the pattern of blind spots, and the reasons for their maintenance in governance systems are bound to be complex. Looking ahead to later chapters, and questions of imagined futures and of community therapy, this complexity has the simple implication of not allowing for a comprehensive mapping of blind spots. The same complexity has the less simple effect of multiplying the effects of blind spots in governance systems, and the pathways of their sometimes-symptomatic formation.

Coping mechanisms for communities can rely on blind spots. Unconscious histories, Master Signifiers, desires, anxieties, exclusions, and internal and external relations can form through blind spots in observation, but they can also be repressed through blind spots (Caruth, 2016). Governance can stabilize blind spots for a while through its discursive and organizational activities, through ensconcing narratives and their blind spots. Blind narratives can be taught at schools, turned into policy or law, and more perspicacious ones can be officially frowned upon or rendered illegal (Foucault, 2012). Governance system will differ in their power to manage blind spots, yet naturalization of contingent realities is a core function of governance, underpinning its task of stabilization of the community, of the relations between individual, community, and environment. Governance needs to simplify its own understanding of the world to function, and it helps people navigate the world in the same movement, reducing the variety of interpretations that are available and normal, in collaboration with other social systems amenable to its instructions to varying degrees (Valentinov, 2014). Governance assists the community in thinking and not thinking.

When terrible things have happened within and to the community, the need for coping mechanisms is more outspoken, and the need not to think, not to see, and to forget some histories, and to encode or impose the memory of others is stronger. Trauma causes gaps in observation and self-observation, but also less dramatic circumstances, that is, seemingly minor conflicts and disagreements, can lead to a proliferation of blind spots (Bistoen, 2016). These can function as part of coping strategies likely to include other mechanisms, such as ritualization and commemoration.

Compromises with neighbors or higher authorities, losses in war, but also difficult internal or external negotiations, can be felt as traumatic. Difficulties in speaking of micro-histories in the community, are not only the difficulties of factions forced to accommodate and accept what was and remains unpalatable. Communities bear the mark of moments of internal division, when grounding mythologies are tested, when it appeared that actual functioning does not coincide with the image of self (Frosh, 2001; Glynos, 2011). What was painful can thus be remembered as such, or hidden from remembering, by factions and community alike. What caused the pain, what caused the deformation of memory can be, for an outsider, something seemingly minor. A compromise can be perfectly workable, yet the need for a compromise might already be felt as hurtful, as a reminder of differences not supposed to exist (Florence, 2021).

Blind spots can also reflect deep *adaptations* to an environment that might not be entirely understood or observed (Luhmann, 1989; Van Assche, Duineveld et al., 2022). Communities change their environments, but this is never a complete change, and never a change that is understood in all its ramifications, as the social-ecological systems always retain hidden corners and invisible mechanisms. Intentional and unintentional blind spots can be involved, where the turning away from things in the environment can, in itself, constitute an adaptation (Duineveld et al., 2017). Elsewhere, it can more realistically be interpreted as an indirect result of adaptations. We have to remind ourselves that adaptations always lead to a set of distinctions that obscure features of the environment. Deep adaptations are not necessarily hidden, as they can be thematized in governance and public discourse. Yet, as they are central to the functioning of the community in a particular environment, because they reflect key choices made in its history, it is more than likely that such deep adaptations are present in both the conscious and the unconscious, and that they are associated with both conscious and unconscious Master Signifiers (Banaji et al., 1997; Hillier & Gunder, 2003).

When the imperfection of a deep adaptation comes to light, the response is likely to be conditioned by more than logic. If a community identifies itself through its taming of the wilderness, its fight against the water, or, alternatively, through its peaceful and mutually beneficial coexistence with nature, and then disaster comes, it is very likely that the community will find ways to explain events, to attempt to brush them away. Questioning core values, and exposing fundamental fantasies is less likely (Žižek, 2008).

Blind spots, therefore, can ignore negatives or positives, as discourse emphasizes negatives or positives, threats, or opportunities. Collective defenses can be the cause when actions, events, or affects do not fit the preferred self-image. It might be helpful to note that blind spots of different ages coexist, that have been caused in histories preoccupied with different themes. These qualitatively different blind spots can, however, interact, create responses, in public discourse, in public policy, which are overdetermined in this specific sense too. The mix of transparency and opacity existing behind any decision, behind anything voiced in governance, is almost always the result of an interplay of blind spots of different sorts, with alternating actors and institutions, stories, and forms of expertise foregrounding new patterns of light and dark all the time. Leaning toward a policy tool reveals blind spots before decisions are taken and will introduce different blind spots through encoding and implementation.

Fantasy and reality testing

What is real is what is felt as real and what is desired (to be real). Fantasy accompanies everything we do and think. Stories structure fantasy, while fantasy generates new stories. These elementary lessons from psychoanalysis have to stay with us when considering the realities of governance, and the testing of those realities (Freud, 1938, 1989, 2015, 2019, 2020). As stories connect to others, so do fantasies, and, indeed, desires. Reality, therefore, is not only a construct but also an inherently unstable construct, while governance has powers of stabilizing realities but does not escape the destabilizing tendency of discourse and desire.

Fantasy scenarios can be highly inventive, and they can borrow from the most trivial and commonplace of narrative structures, figures, and tropes (Žižek, 1992, 2019). If we allow for positive and negative fantasy, we can say that fantasy can stage desire or fear, or both. What we fear is tinged by what we desire, and what we desire to be. What is presented as reality in governance, is structured by narrative anyway, so the narrative elements that are associated with reinforcement or the undermining of identity are likely to find their way into the collective discourse about important decisions (Benwell, 2006; McGowan, 2017). Even if there are no obvious similarities between narrative fragments in political or administrative rhetoric and what plays out in collective desire, fantasy is at work, supporting the affective investment required for policies to have effect, and filling in the blanks where logical connections are vague at best (Žižek, 2006).

Desire needs to be maintained, as what is more atrocious than facing what is hated, is the dissolution of desire and the vanishing of affect, rendering life meaningless. A truly unfortunate side effect of administrative routines, invented to make practical governance easier, can be exactly that. What might have been the result of passionate debate, what might have roused the collective, turns into mindless hairsplitting and an incapacity of the administrative system to communicate about goals and assumptions, let alone emotion.

What is staged in the fantasy is the functioning of desire itself, a graspable way for the subject to represent its desire to itself, and to maintain it (Frosh, 2001). Indeed, each community is a productive fiction, in the sense that it comes into being through narrative, a narrative that can never perfectly capture the sense of reality, the set of identifications, the desire of individuals (Žižek, 1992). The grounding fantasy of community therefore serves to catalyze the formation of other fantasies, the expression of other desires – as now there is something that can desire, that can strive toward a better version of itself, toward a better future (Bistoen, 2016; Vanheule & Verhaeghe, 2009).

The reality for a community, for a governance system trying to improve it according to the desires of citizens is thus always starting from a myth, a story driven by the desire to be a community. If it is imposed, one can hardly speak of a community but rather of a group of individuals being placed in a political order. Over time, a community might form, Voice might emerge through governance, and a new myth be created, which might differ significantly from the one in the books of the ruling ideology.

Against this background, it might appear as natural that reality testing in governance cannot be a simple matter. We reformulate insights from this and previous chapters as three reasons for the difficulties in conceptualizing and operationalizing reality testing. Presenting reality testing, and its derivative, the assessment of success and failure, as a matter of simple observation, stems from the needs of simplifying ideologies.

One difficulty is that, as we know, reality is not something that can be accessed directly and known objectively, independently of the world of individual histories and collective culture. For Freud, the reality principle did not presume the presence of a reality that can be easily accessed and compared with desired states (Freud, 2003, 2014). He invoked it as a limit, an obstacle for the pleasure principle. Individuals cannot pursue their own desires under all conditions and cannot do this without hitting limitations. The reality delineated here is thus one of multiple barriers, often not fully understood. The hindering involved can be a matter of material, psychological, or social realities infringing upon the free pursuit of pleasure (Van Haute, 2002). Those realities do cohere for the individual, but this does not mean all barriers are transparent, nor that the same coherent pattern holds under all conditions, and it still requires fantasy to stitch together the fragments experienced or observed (Catlaw & Marshall, 2018).

A second reason is that the construction of the reality that is used as the norm for the reality testing, is still a construction, affected by the histories of those involved in the construction, by affect, desire, and the co-evolution of discourse (Foucault, 2001, 2003, 2010). In governance, the situation can be more complex, as the dominant discourses in governance become involved in both reality construction and reality testing (Van Assche et al., 2023). This can be the case through the assessment of success or the failure of policies, but also through everyday discussions, through interpretation of signals from society and the delineating of problems and preference for methods of investigation and testing. Reality building and reality testing are thus almost always entwined in governance, and the exact nature of that entwining is not easy to observe. One can say it requires second-order observation.

Third, there is real value in stressing the limited value of adaptation to reality. Psychoanalysis, in the tradition of Freud and Lacan, does not assume that adaptation to reality is a goal (Van Haute, 2002). Nor is there a model of a perfectly balanced or integrated personality, for which adaptation to reality, to society, should not be too much of a burden. Psychoanalysis is not in the business of normalizing or moralizing. Testing, as in comparing to a norm of being or of adapting and fitting into a context, is therefore not an overriding goal for individuals. Nor should it be for communities, and if governance systems and experts contribute to a narrowing down of normality and to Superego injunctions to adapt to that reality, this is a problem (Frosh, 2016; McGowan, 2017). Roland Barthes already noticed that what was self-evident and natural for his mainstream media and his government, was the ideology of the bourgeoisie carefully veiling its own grounding mythologies (Barthes, 1957; Iversen, 2007). Governance systems endowed with powers to shape society and its realities therefore cannot be left alone uncritically when using a rhetoric of adaptation and normality. Policy practitioners, experts in economics and law, are not to be confused with people closer to a reality worth adapting to; they create a reality that fits their categories and interests, and then reflects their experiences.

The testing of policy outcomes, but also the more basic testing of realities assumed and built-in governance, will involve both assumptions on reality and norms for adjustment to that reality. Success, whether defined in generic evaluative terms, quantitative terms, or in terms of conformance with predefined goals, will look and feel different in the community (Beunen et al., 2013). Reality testing in governance is thus always a testing of desire, a serious exercise in self-reflection, trying to grasp whether the series of translations of external signals (from the community) into problem definition and solution, into policy formation and implementation, alienates the community, ends up denying the truth of its desire. It cannot be easy, as it requires an interrogation of the desires driving the governance system, and of the fantasies structuring the calls for stability or change in the community (Dunlop, 2020; Van Assche et al., 2012). Confronting the community with some of the more problematic aspects of such fantasies in collective discourse and identification is by definition prickly, as is the case in democracies where a common assumption in governance is that leadership represents only or exclusively what people think.

This is especially true when the fundamental fantasy underpinning community identity deserves questioning, either because of conflicts internal and external, or because of severe self-limitation, which is bound to create problems later (Gunder, 2021; Žižek, 2009). If what is questioned is not merely a fantasy about the future or one aspect of policy but a fundamental fantasy of community identity that coordinates key values and associated Master Signifiers, then asking the community to rethink itself, and thereby expose its identity as a fiction that has lost its productivity, is inherently risky for leadership and painful for the rest (Foroughi, 2020; Gardner et al., 2021).

Governance entails the naturalization of what is contingent. It can be performative, in the sense that what was fantasy, can become reality. The process of naturalization can increase rigidity, meaning that an envisioned reality becomes experienced reality for many, and less easy to adapt to. Adaptation tends to reduce adaptive capacity, and the loss of friction between the realities of governance and community can therefore come with serious drawbacks. Hence, the difficult oscillation in the life of a community between embracing and questioning fantasy, between not questioning modes of thinking, organizing, and of their connections in governance, and episodes and sites of diversifying perspectives and enhancing critical self-analysis (Van Assche et al., 2024). The hardening of realities in and through governance serves to stabilize individuals and communities in their environment yet renders these realities brittle.

Concluding

Governance systems do not exist independently from the communities they represent and organize, and they do not remain unaffected by the realities constructed in those communities. What is real is what is *felt* to be real and what is desired; governance systems interfere in both aspects. Blind spots cannot be avoided, governance creates its own realities, and the relations between governance, community, and environment make it impossible to objectively test the realities present in governance and therefore of policy success or failure. Democratic innovation that perceives delusions and abuse in current governance practices, is not immune to blind spots itself, and at the same time, the need remains for supportive fantasies that gloss over cracks in discourse and affect. Governance systems can learn, meaning learn about reality, but also about a construction of realities, an entangling that remains unproblematic as long as governance can continue, and as long as those realities are accepted.

For Voice to emerge, shared fantasy must support shared discourse. Voice can amplify the power of governance to create reality effects, to transform discursive and material worlds. Governance can trip over the Real in diverse ways, just as the visions underpinning policies can become unproductive fantasies through many paths. Reconstructing those paths becomes more interesting when there is a desire to do so, when there is a collective recognition of a problem, and a willingness to reflect on old policy tools, goals, and their grounding narratives (we refer to our last chapter for the implications in terms of community therapy). For an alternative understanding of reality testing, we now briefly turn to Russia.

Russian cultural theorist Mikhail Bakhtin coined the term *polyphony* for cultural products where a unified voice is present, while diverse identities maintain an internal coherence – as in his analyses of Fyodor Dostoevsky's novels (Hirschkop & Shepherd, 2001). Polyphony in this sense is a normative concept, a style Bakhtin favored, a form of literature (cultural production) that allows to represent and grapple with the complexity of life and society in a more sophisticated manner. At the same time, polyphony appears with him as a non-normative feature of language, of discourse, of all cultural production, entailing all sign systems: every utterance is part of a world of other utterances that only functions because of the continuous fusion of voices (Holloway & Kneale, 2000). A story, a speech, a strategic plan only means something, and only functions because it rests on a history of other speeches, stories, and plans, in a structure where other stories resonate. Other genres, other perspectives, other cultural forms and media all contribute to the semiotic melting pot that makes both interpretation and creativity possible.

In the spirit of polyphony, we can come to understand the activity of reality testing in governance as ideally relying neither on monolithic expert knowledge nor on the proliferation of individual opinions and local sentiment. Opening a presumably closed system to "the community," supposedly excluded before, does not resolve the issue of reality testing and we need to lean on Bakhtin both in the normative and the analytic sense. His descriptive notion of polyphony easily fits into the psychoanalytic tradition and in our unfolding perspective on governance, his normative idea helps retain an ideal of democracy around a concept of unity in diversity (Roberts, 2004). More precisely, the testing of reality that is required in any modern democracy can peacefully coexist with the acknowledgment of the tenuous existence of a collective voice, and the acknowledgment of the ultimately narrative and fictitious character of both the unified and diverse voices (Capizzo, 2018).

Tracing the analysis of realities in governance to observation helps in better mapping governance but does not lead us to more solid ground. Psychoanalysis demonstrated that an observer comes to observe herself and the world through images derived from elsewhere, influenced by contingent events, and that a history of observations shapes future observation. To grasp a broader set of limitations and structuring principles for reality and reality testing in governance, we therefore have to shift our own observation to the world of emotions, individual and collective, which is what we will do in the following chapter.

References

Alvesson, M., & Spicer, A. (2016). *The Stupidity Paradox: The Power and Pitfalls of Functional Stupidity at Work*. Profile Books.

Bal, M. (2002). *Travelling Concepts in the Humanities: A Rough Guide*. University of Toronto Press.

Banaji, M. R., Blair, I. V., & Glaser, J. (1997). Environments and Unconscious Processes. In *The Automaticity of Everyday Life*. Psychology Press.

Barthes, R. (1957). *Mythologies*. Éditions du Seuil.

Benwell, B. (2006). *Discourse and Identity*. Edinburgh University Press.

Beunen, R., Van Assche, K., & Duineveld, M. (2013). Performing failure in conservation policy: The implementation of European Union directives in the Netherlands. *Land Use Policy*, *31*, 280–288. https://doi.org/10.1016/j.landusepol.2012.07.009

Bistoen, G. (2016). The Lacanian Concept of the Real in Relation to Politics and Collective Trauma. In G. Bistoen (Ed.), *Trauma, Ethics and the Political beyond PTSD: The Dislocations of the Real* (pp. 104–130). Palgrave Macmillan UK. https://doi.org/10.1057/9781137500854_6

Capizzo, L. (2018). Reimagining dialogue in public relations: Bakhtin and open dialogue in the public sphere. *Public Relations Review*, *44*(4), 523–532. https://doi.org/10.1016/j.pubrev.2018.07.007

Carr, E. S. (2010). Enactments of expertise. *Annual Review of Anthropology*, *39*, 17–32. https://doi.org/10.1146/annurev.anthro.012809.104948

Caruth, C. (2016). *Unclaimed Experience: Trauma, Narrative, and History*. JHU Press.

Catlaw, T. J., & Marshall, G. S. (2018). Enjoy your work! The fantasy of the neoliberal workplace and its consequences for the entrepreneurial subject. *Administrative Theory & Praxis, 40*(2), 99–118. https://doi.org/10.1080/10841806.2018.1454241

Clegg, S., Biesenthal, C., Sankaran, S., & Pollack, J. (2017). Power and Sensemaking in Megaprojects. In B. Flyvbjerg (Ed.), *The Oxford Handbook of Megaproject Management* (p. 0). Oxford University Press. https://doi.org/10.1093/oxfordhb/9780198732242.013.9

Duineveld, M., Van Assche, K., & Beunen, R. (2017). Re-conceptualising political landscapes after the material turn: A typology of material events. *Landscape Research, 42*(4), 375–384. https://doi.org/10.1080/01426397.2017.1290791

Dunlop, C. (2020). Policy Learning and Policy Failure: Definitions, Dimensions and Intersections. In C. Dunlop (Ed.), *Policy Learning and Policy Failure* (pp. 1–22). Bristol University Press. https://doi.org/10.46692/9781447352013.001

Eco, U. (1979). *The Role of the Reader: Explorations in the Semiotics of Texts*. Indiana University Press.

Esposito, E. (2008), Social forgetting: A systems-theory approach. In A. Erll & A. Running, (Eds.), Cultural memory studies. De Gruyter.

Ferguson, J. (1994). *The Anti-Politics Machine: "Development", Depolicization, and Bureaucratic Power in Lesotho*. University of Minnesota Press.

Fink, B. (1999). *A Clinical Introduction to Lacanian Psychoanalysis: Theory and Technique*. Harvard University Press.

Fischer, F. (2000). *Citizens, Experts, and the Environment: The Politics of Local Knowledge*. Duke University Press.

Florence, J. (2021). *Identification in Psychoanalysis: A Comprehensive Introduction*. Routledge. https://doi.org/10.4324/9781003154426

Foroughi, H. (2020). Collective Memories as a Vehicle of Fantasy and Identification: Founding Stories Retold. *Organization Studies, 41*(10), 1347–1367. https://doi.org/10.1177/0170840619844286

Foucault, M. (2001). *Madness and Civilization: A History of Insanity in the Age of Reason*. Routledge.

Foucault, M. (2003). *Society Must Be Defended: Lectures at the College de France, 1975-76*. Allen Lane The Penguin Press.

Foucault, M. (2010). *The Birth of the Clinic: An Archaeology of Medical Perception* (1. publ., reprinted). Routledge.

Foucault, M. (2012). *Discipline and Punish: The Birth of the Prison*. Knopf Doubleday Publishing Group.

Freud, S. (1938). Constructions in analysis. *The International Journal of Psycho-Analysis, 19*, 377.

Freud, S. (1989). *The Ego and the Id* (J. Strachey, Ed.). Norton.

Freud, S. (2003). *The Uncanny* (H. Haughton, Ed.; D. McLintock, Trans.; Illustrated edition, pp 123–162). Penguin Classics.

Freud, S. (2014). *The Neuro-Psychoses of Defence*. Read Books Ltd.

Freud, S. (2015). *Beyond the Pleasure Principle* (J. Miller & M. C. Waldrep, Eds.). Dover Publications.

Freud, S. (2019). I. Creative Writers and Daydreaming. In E. Kurzweil & W. Phillips (Eds.), *I. Creative Writers and Daydreaming* (pp. 19–28). Columbia University Press. https://doi.org/10.7312/kurz91842-003

Freud, S. (2020). *The Interpretation of Dreams: The Psychology Classic*. John Wiley & Sons.

Frosh, S. (2001). Psychoanalysis, identity and citizenship. In N. Stevenson (Ed.), *Culture and Citizenship*, 62–73.

Frosh, S. (2013). Psychoanalysis, colonialism, racism. *Journal of Theoretical and Philosophical Psychology*, *33*(3), 141–154. https://doi.org/10.1037/a0033398

Frosh, S. (2016). Relationality in a time of surveillance: Narcissism, melancholia, paranoia. *Subjectivity*, *9*(1), 1–16. https://doi.org/10.1057/sub.2015.19

Gardner, W. L., Karam, E. P., Alvesson, M., & Einola, K. (2021). Authentic leadership theory: The case for and against. *The Leadership Quarterly*, *32*(6), 101495. https://doi.org/10.1016/j.leaqua.2021.101495

Glynos, J. (2011). On the ideological and political significance of fantasy in the organization of work. *Psychoanalysis, Culture & Society*, *16*(4), 373–393. https://doi.org/10.1057/pcs.2010.34

Gunder, M. (2005). Lacan, planning and urban policy formation. *Urban Policy and Research*, *23*(1), 87–107. https://doi.org/10.1080/0811114042000335287

Gunder, M. (2021). Fantasy in planning organisations and their agency: The promise of being at home in the world. *Urban Policy and Research*, *39*(3), 218–232. https://doi.org/10.1080/08111146.2021.1989776

Gunder, M., & Hillier, J. (2009). *Planning in Ten Words or Less: A Lacanian Entanglement with Spatial Planning*. Ashgate Publishing, Ltd.

Healey, P. (2004). Creativity and Urban Governance. *disP - The Planning Review*, *40*(158), 11–20. https://doi.org/10.1080/02513625.2004.10556888

Hillier, J., & Gunder, M. (2003). Planning fantasies? An exploration of a potential Lacanian framework for understanding development assessment planning. *Planning Theory*, *2*(3), 225–248. https://doi.org/10.1177/147309520323005

Hirschkop, K., & Shepherd, D. (2001). *Bakhtin and Cultural Theory*. Manchester University Press.

Holloway, J., & Kneale, J. (2000). Mikhail Bakhtin: Dialogics of space. In M. Crang, & N. Thrift (Eds.), *Thinking Space*. Routledge.

Hood, C., & Peters, G. (2004). The middle aging of new public management: Into the age of paradox? *Journal of Public Administration Research and Theory*, *14*(3), 267–282.

Howe, K. R. (2004). A critique of experimentalism. *Qualitative Inquiry*, *10*(1), 42–61. https://doi.org/10.1177/1077800403259491

Iversen, M. (2007). *Beyond Pleasure: Freud, Lacan, Barthes*. Penn State Press.

Johnston, A. (2005). *Time Driven: Metapsychology and the Splitting of the Drive*. Northwestern University Press.

Kim, E., Zeppenfeld, V., & Cohen, D. (2013). Sublimation, culture, and creativity. *Journal of Personality and Social Psychology*, *105*(4), 639–666. https://doi.org/10.1037/a0033487

Lasch, C. (1992). The Culture of Narcissism. In R Wilkinson (Ed.), *American Social Character*. Routledge.

Latour, B. (2004). *Politics of Nature: How to Bring the Sciences into Democracy*. Harvard University Press. https://doi.org/10.4159/9780674039964

Luhmann, N. (1989). *Ecological Communication*. University of Chicago Press.

Luhmann, N. (1995a). *Social systems* (Vol. 1). Stanford University Press Stanford.

Luhmann, N. (1995b). The paradoxy of observing systems. *Cultural Critique*, *31*, 37–55. https://doi.org/10.2307/1354444

McAfee, N. (2022). Public Philosophy and Deliberative Practices. In *A Companion to Public Philosophy* (pp. 134–142). John Wiley & Sons, Ltd. https://doi.org/10.1002/9781119635253.ch14

McGowan, T. (2017). *Capitalism and Desire: The Psychic Cost of Free Markets.* Columbia University Press. https://doi.org/10.7312/mcgo17872

Okulicz-Kozaryn, A., & Valente, R. R. (2018). City life: Glorification, desire, and the unconscious size fetish. *Psychoanalysis and the GlObal.*

Roberts, J. M. (2004). From populist to political dialogue in the public sphere: A Bakhtinian approach to understanding a place for radical utterances in London, 1684–1812. *Cultural Studies, 18*(6), 884–910. https://doi.org/10.1080/0950238042000306918

Rochefort, D. A., & Cobb, R. W. (1994). *The Politics of Problem Definition: Shaping the Policy Agenda.* University Press of Kansas.

Ruti, M. (2008). Why there is always a future in the future. *Angelaki, 13*(1), 113–126. https://doi.org/10.1080/09697250802156109

Ruti, M. (2012). *The Singularity of Being: Lacan and the Immortal Within.* Fordham University Press.

Scott, J. C. (2020). *Seeing Like a State: How Certain Schemes to Improve the Human Condition Have Failed.* Yale University Press.

Strassheim, H., & Kettunen, P. (2014). When does evidence-based policy turn into policy-based evidence? Configurations, contexts and mechanisms. *Evidence & Policy, 10*(2), 259–277. https://doi.org/10.1332/174426514X13990433991320

Tingle, N. (1992). Self and liberatory pedagogy: Transforming narcissism. *Journal of Advanced Composition, 12*(1), 75–89.

Valentinov, V. (2014). The complexity–sustainability trade-off in Niklas Luhmann's social systems theory. *Systems Research and Behavioral Science, 31*(1), 14–22.

Van Assche, K., Beunen, R., & Duineveld, M. (2012). Performing Success and Failure in Governance: Dutch Planning Experiences. *Public Administration, 90*(3), 567–581. https://doi.org/10.1111/j.1467-9299.2011.01972.x

Van Assche, K., Beunen, R., & Gruezmacher, M. (2024). *Strategy for Sustainability Transitions: Governance, Community and Environment.* Edward Elgar Publishing.

Van Assche, K., Beunen, R., & Verweij, S. (2020). Comparative planning research, learning, and governance: The benefits and limitations of learning policy by comparison. *Urban Planning, 5*(1), 11–21.

Van Assche, K., Duineveld, M., Beunen, R., Valentinov, V., & Gruezmacher, M. (2022). Material dependencies: Hidden underpinnings of sustainability transitions. *Journal of Environmental Policy & Planning, 24*(3), 281–296. https://doi.org/10.1080/1523908X.2022.2049715

Van Assche, K., Gruezmacher, M., & Beunen, R. (2022). Shock and conflict in social-ecological systems: Implications for environmental governance. *Sustainability, 14*(2), 610. https://doi.org/10.3390/su14020610

Van Assche, K., Gruezmacher, M., Marais, L., & Perez-Sindin, X. (2023). *Resource Communities: Past Legacies and Future Pathways.* Taylor & Francis.

Van Haute, P. (2002). *Against adaptation: Lacan's "subversion" of the subject.* Other Press.

Vanheule, S. (2016). Capitalist discourse, subjectivity and Lacanian psychoanalysis. *Frontiers in Psychology, 7*. https://doi.org/10.3389/fpsyg.2016.01948

Vanheule, S., & Verhaeghe, P. (2009). Identity through a psychoanalytic looking glass. *Theory & Psychology, 19*(3), 391–411. https://doi.org/10.1177/0959354309104160

Voß, J.-P., & Bornemann, B. (2011). The politics of reflexive governance: Challenges for designing adaptive management and transition management. *Ecology and Society, 16*, 9.

Wagenaar, H. (2015). 22 Transforming perspectives: The critical functions of interpretive policy analysis. In F. Fischer, D. Torgerson, A. Durnová, & M. Orsini (Eds.), *Handbook of Critical Policy Studies* (pp 422–440). Edward Elgar.

Žižek, S. (1992). *Looking Awry: An Introduction to Jacques Lacan through Popular Culture*. MIT Press.

Žižek, S. (2006). *Interrogating the Real*. Bloomsbury Publishing.

Žižek, S. (2008). *Violence*. Picador.

Žižek, S. (2009). *In Defense of Lost Causes*. Verso. http://public.ebookcentral.proqu est.com/choice/publicfullrecord.aspx?p=5176960

Žižek, S. (2019). *The Sublime Object of Ideology*. Verso Books.

5 Affect and governance

What appears to be rational, is rarely only that. Freud discovered this early on, admitting freely that his ideas were foreshadowed by the explorations of poets and novelists. Later in life, he devoted much time to mapping out the interconnections between individual and collective affect, often tracing them back in deep history (Freud, 2004a, 2004b, 2015). People feel things because of other people, because they learn to recognize and categorize affect through interactions with others. Feelings are explained in shared languages and cultures, and serve to explain actions, ideas, and other feelings. People care about identities, and identities shape the affective economy.

If, for modernism, governance must be driven by reason, and if reason and emotion can be clearly distinguished, for psychoanalysis, this is not the case. In governance, moreover, the entwining of reason and affect has specific features that can be mobilized by those aspiring to power, and those in power. Both politics and administration, sometimes in tandem, can avail themselves of strategies of intellectualizing, rationalizing, and of excessive cathexis. Emotion can be hidden or emphasized, used and abused, transferred or transformed. Such strategies can mobilize support, and can convince people of proposed changes or of the reality or desirability of the world assumed in policies and procedures.

Narratives allow us to recognize affect and meaning, to make a meaning truly felt. Psychoanalysis discovered a century ago that what we feel, what we want, fear, and hope, greatly affects how we construct our realities (Freud, 2003, 2020, 2023). Stories play a mediating role, as the selection, modification, interpretation, and performance of stories helps people manage feelings, partly by making them real. Once a new narrative is in the world, it can be connected to other narratives, places, and topics in ways unexpected by those crafting the narrative, and this entails that new affects can be discovered in and through them (Eco, 1979). Through the same paths, new affects can become recognizable: a very particular worry about something, a new shade of nostalgia, a new modality of hope. New stories, songs, and movies can give expression to the new affect, which can in turn reframe other stories,

DOI: 10.4324/9781032696676-5

and establish new connections between previously recognized emotions (Green, 1999).

Emotions are often presented as more real than words and actions, closer to identity, even for those who otherwise understand reality as primarily accessible through reason. Good leadership is then authentic leadership, recognizing emotional intelligence, and good governance entails direct participation, where the emotions of stakeholders can be expressed, explored, and discussed (Alvesson & Einola, 2019). For psychoanalysis, however, authenticity is not a helpful notion, as we all construct ourselves by means of foreign elements, discursively and affectively (Fink, 1999; Knudsen et al., 2016). One could even argue, following Slavoj Žižek and Lacan (Žižek, 1992, 2011), that what is most intimate, most confusing, most difficult to manage for the individual, relies more on the collective, on conventional, even ritualized signifiers. In ritual, meaning is expected to emerge from codification, repetition, and display rather than from feeling *or* understanding.

Clinging to a feeling might be more important than anything, and communities might sacrifice much, including a workable sense of reality and much enjoyment in community life, in order to maintain a feeling, which might keep other feelings at bay. Fear of chaos, hope for a well-ordered world, desire for inclusive social bonds, and anxiety about the environment can keep a community together, and questioning them can be most unwelcome, or welcomed with a dose of aggression. A sense of self and a stable reality might be inferred from an affect or sometimes from a desire for something to be true, for a new or existing identity, and sometimes more negatively, from a desire to not-know and not-feel. None of this, however, means that affect is primary, neither in an ontological sense nor in the sense of rhetorical superiority nor because of a presumed authenticity (Fink, 1999, 2013; Soler, 2015).

Not feeling and not telling

In governance, there are good reasons for saying yet not feeling. As there are many occasions for feeling but not saying. This leaves ample space for rhetoric, for performance of affect, and for stories laden with affect (Roelvink, 2010; Salvatore et al., 2021). Stories of tradition, core values, and of belonging are commonly deployed to manage public affect, while hope and fear are always present whenever the future is concerned. Moreover, what a community is and what it desires to be, cannot be separated (Clarke, 2005). Communities do not operate on the basis of objectively existing needs. Good governance caters to the desires of a community. What counts as luxury, or even as non-existent, can appear as a unifying goal of public policy elsewhere, as a sine qua non for anyone in governance to survive (McGowan, 2022).

Thus reasons not to show emotion, and reasons to fake emotion abound in governance (Hunter, 2015). Arousing emotions and respecting emotions do not require sharing those emotions. Real emotions, about the community, a policy, an opponent, might better be expertly veiled. A distinct reason for non-articulation of what is felt, and pretending to feel something by those in governance, beyond the need to please and work together, is that what the community wants, might require things people do not want or do not want to see. As Otto von Bismarck noted, nobody wants to know what goes into the sausage, while Machiavelli observed that transparency does not serve those who serve the public (McCormick, 2011). The implications of and requirements of border closure, mega-projects, of merging communities, might not be clear, and people might not be interested in knowing them (Flyvbjerg, 2001, 2004). A measure of feint, even deceit, might be part of the job for all interested in participation in governance. Changing or hiding stories, interpretations, or facts can be order of the day.

Another reason for dissimulation stems from a different practical limit of governance. As governance is the art of the possible, is often the art of compromise, this means that actors rarely get what they want. They can rarely maintain a story for a long period of time, as what was promised cannot be delivered, but also in the sense that the sharpness and sweetness of early rhetoric cannot be maintained. Things might not be as rotten as we presented them, some goals might be impossible to achieve, some administrative functions might be more necessary or costly than imagined, and other parties might not be as evil as older speeches might have it. The need to collaborate in order to make things work, the proximity of other stories, and the discursive creativity of governance mean that stories and actions will diverge, and that stories are bound to lose their initial purity, a purity that might have been effective in mobilizing public emotion (Grube, 2014). When that purity is lost, people can feel betrayed.

Careful rhetoric and cautious management of one's emotional performance have thus been part and parcel of governance since the Stone Age (Consigny, 1974). Politics does not reward emotional authenticity, but neither does it benefit those unable to display any emotion. This applies especially to those actors who are most visible to the community, those who have made the most unambiguous promises, and those who cannot evade responsibility. Spreading responsibility, through elaborate participation processes or through bureaucratic specialization, does not always solve the problem, as success can still be achieved through emotional rhetoric and clear-cut narratives, giving space to new competitors willing to use them, and as unclear responsibility can increase public desire to designate a scapegoat (Bovens, 1998; Drabeck & Quarantelli, 2008).

Strategies of intellectualizing, rationalizing, and of excessive cathexis

In governance, different players have different reasons to emphasize emotion or reason, and different strategic options for relating and performing reason and affect. This is not so strange, as the relations with other actors and with the community, and the prevalent ideas and affects pertinent to the issue can all vary widely. *Intellectualizing* what was in essence a decision or reasoning driven by affective, political, or otherwise non-expert motivations, is a common way to relate reason and affect. It appears most common where expert knowledge is appreciated and considered non-controversial (Fischer, 2009; Latour, 2004). Where expert knowledge imbues a policy or a decision-maker with prestige, this can lead to the proliferation of expert knowledge in administration, and to the self-styling of politicians as intellectuals. It might open possibilities for universities to exert influence, or for technically-oriented consultants to provide expertise.

Intellectualizing can assist politicians to evade responsibility, in either avoiding taking decisions (now left to experts) or using expertise as a smoke screen for decisions taken for other reasons. It can help administration focus on the reproduction of the bureaucratic systems of rules, roles, and procedures, while avoiding a focus on politically and culturally sensitive issues (Ferguson, 1994; Scott, 2020). Intellectualizing can thus, paradoxically, contribute to functional stupidity in organization, as testified by Mats Alvesson and Andre Spicer's analyses of the proliferation of *fachidioten* in organizations fostering non-thinking (Alvesson & Spicer, 2016). When trust in experts and expertise declines, then most certainly fewer actors are inclined to emphasize expertise, to build up elaborate reasoning connecting policy issues to expert knowledge (Fieschi & Heywood, 2004; Merkley, 2020).

Fear can be involved in intellectualizing: fear of confrontation with emotion, fear of conflict, of direct interaction with different viewpoints. Real difference is de-mined through intellectual discourse. The conflict avoided can be with the public waiting outside, or internal, as when competing identifications and desires need to remain separated or suppressed (Gould et al., 2013). Alternatively, not fear but desire for, or a hopeful belief in the existence of a well-ordered world can underlie the habit of intellectualizing.

Translation of desire into need can use the shortcut of intellectual categories, while avoiding meaningful dialogue. In a polyphonic environment, even when voices are acknowledged, they want to be seen as distinct, and their interplay is bound to rules of composition. In other words, a splitting of the world of voices into "actors," "stakeholders," "resources," "uses," and "users," as well as "preferences" helps simplify the world for those in governance, to recognize clearly delineated groups with clear perspectives, objectively definable needs, and sometimes demands that might exceed those needs and the limits of existing resources (R. Hulme & Hulme, 2012; Jun, 2012; Newman & Clarke, 2009). "The people" want a variety of things, demand them, might not know their own desires, but governance must respond to this chaos.

Rationalizing is a more elementary strategy of stabilizing emotions and interactions. In governance, rationalizing can lead to intellectualizing, but not necessarily so (cf. Zepf, 2011). Rationalizing can happen without reference to expert knowledge, without developing elaborate argumentation, and it can occur without splitting emotions from arguments. It can simply be a move from a feeling to an argument, a public appeal to logic where something else prevailed. The logic can be a public logic where perhaps private reasons were at work, or a private logic where affect was driving the pronouncement or proposal. As with intellectualizing, rationalizing can be strategic, cultural, or otherwise. "Otherwise" can refer to the blunt pursuit of power in and through governance, it can be the pursuit of collective goals through means not appreciated by the collective, and it can be particular desires, conscious or unconscious, that might even benefit the community.

As intellectualizing, rationalizing can tamp down conflict, create unified discourse, stabilize the relation between governance and community. It can help to find consensus and fake consensus (Tewdwr-Jones & Allmendinger, 1998). It can help to work through prickly issues and to avoid those issues. Rationalization will certainly benefit those operating in an environment steeped in a belief in rational solutions, in the capacity of reason to give us access to an objective problem in an epistemically unified world (Allmendinger, 2017).

Rationalization can thus take on the form of bureaucratization, where the world is understood through papers and forms, and whatever does not fit the paperwork is dismissed as irrelevant or nonexistent. This is the world of Franz Kafka and absurdist comedy, but a world very real for those who spend time in large organizations, both public and private (Alvesson, 1987; Fineman, 2000). People can come to identify with a role, a procedure, a rule, or, in old Soviet style, a stamp. The stamp comes to represent an identity, an ordered world, but also the power to order that world. Doubting the reality of the paperwork is doubting the capacity of those papers to order the world, if not to solve actual problems, then at least to keep chaos at bay (Åkerström et al., 2021).

Bureaucratization makes it difficult for alternative understandings of the world to enter the governance system, and it makes it difficult to reorganize governance in ways not understood through current paperwork (Anderson, 2021). On the other hand, the underlying belief in the power of bureaucracy to shape and improve reality, can endow the system with great power (Clegg et al., 2016). A confident governance system, confident in its tools and its grasp of reality, can achieve more than a system bogged down in reflexivity, plurality, and epistemic doubt. Bureaucratic thinking, thus, cannot be summarily dismissed, yet ideally placed in a governance system that retains spaces for other modes of relating to the world (Luhmann, 2018).

Excessive cathexis can be presented as the inverse of the previous strategies, as it entails an emotional overinvestment in something that can be either rational or affectively driven. The environment dictates what will make sense (Patz et al., 2022). If issues are sensitive in the community, a cool technical approach will not likely be appreciated, and a public affectation not aligned with public sentiment comes at great risk (Neuman, 2007). Machiavelli already noted that political rhetoric in most cases was affective rhetoric (Viroli, 1998), and that argumentation based on facts and logic was not often effective unless those facts fit other stories and their affective framing. Thus, again, stories appear as mediating devices, linking fact and affect, connecting ideas and identities. If anger is considered irrational and weak, or, conversely, if it is felt as a proud and strong reaction to injustice, this will modulate the performance of anger, the appeal to anger, and the interpretation of one's feelings (Mesquita et al., 2017).

The cultural and strategic landscape of governance requires that sometimes, a calm appearance needs to be cultivated, a distance to a real and strong affect, while elsewhere, the performance of affect needs to conceal a calculating stance. Since Freud, we know that calculation and affect can coexist, that it can be hard to distinguish frontstage and backstage in this coexistence (Brennan, 2004). Moreover, the reasons for strong feelings can be entirely opaque to a person, even a very strategically inclined one, while contradictory feelings can (un)happily live together for a very long time (Freud, 1989, 2014, 2020). Excessive cathexis commonly coexists with rationalizing strategies, while competing affects can underlie or accompany a communication or decision (see Chapter 7).

Identity and community revisited

Narratives assist in the recognition, articulation, and arousal of affect, while affect renders narrative persuasive. These effects can take place in the situation, but also later, when other events or narratives establish an association. What people react to in relation to governance is not the narrative per se, not even the policy, but often the perceived effect in everyday life. Those effects, however, are assessed through the lens of narrative (Yanow, 2000). Making sense of policy effects, policy narratives, and of policies themselves, are three different things. Very rarely does a situation occur where leadership is capable of shaping the narrative frames of those observing them so completely that this difference disappears (Wagenaar, 2015).

Identity narratives, we know, are never stable, never unambiguous, but help in the stabilization of processes and relations that would otherwise not crystallize into a unity capable of taking action and orienting itself in the world (Brockmeier, 2002; Holstein & Gubrium, 1999). Not all identity narratives are effective, and not all maintain their power of identification and affective persuasion. People can feel at home in a group, in a place, in a version of history, and then not (Soler, 2015). Identity narratives can aspire to unify a group yet come with such stringent criteria that other values are in peril, and the narrative or the group appear in a negative emotional light: this group is really different, or I do not belong here. In governance, those who identify and those who want others to identify take different positions in a competition of identifications (Rancière, 2023; Žižek, 2002).

The power of identity narratives, their emotional salience and naturalizing power, stems from their capacity to link to other narratives. Some are more encompassing than others, only some narratives assume that identity is malleable through governance, and not all resonate emotionally in daily activities

(Massumi, 2015; Vanheule & Verhaeghe, 2009). Totalitarian regimes attempt to impose unifying and totalizing identities at the service of the state; religion might perfuse everyday life or remain restricted to rituals of unity or transition. If identity narratives entail a closure of the community and a unique claim on identity, coexistence with others, self-questioning, and self-definition all become tenuous undertakings (Benwell, 2006).

Nevertheless, even closed identity narratives, aspiring to shape the affective economy of groups, do not have an eternal grip on actual processes of identification, as they do not control the feral multiplication of interpretation and the promiscuous hybridization of narrative. People will come and go, and both the questions left by those who left, and the fresh eyes of those who arrive make thinking and feeling in the community less predictable. Newcomers see the results of governance and its stories in a new light, and can also find new resonances, positive and negative, in the stories themselves (Tanggaard, 2013).

Fear and hope

No emotion is inaccessible for communities. Groups are capable of hatred and envy, love, passion, and affection, indignation and anger, gratitude, pity, shame, and grief (Freud, 1990, 2004b, 2015). Emotions can spread in groups, through stories, fast or slow, triggering new emotions, recalling and transmuting other stories. Communities create governance systems that transform those communities, alter their affective landscape, and are invested with affect themselves. Mapping out those emotional landscapes is a task for those in governance, and those desiring to participate in governance. The dialectics is always contingent, and the path of governance and community frames the possibilities for new narratives to modify the emotional repertoire.

We are interested in governance for the future, in the possibilities of communities to imagine and organize alternative futures (Van Assche et al., 2023, 2024), and in that perspective, two clusters of emotions acquire exceptional relevance, circling around fear and hope. Fear of disaster, of disappearing from the map, of dissolving into nothingness or irrelevance often, unconsciously, accompanies housekeeping versions of governance, while utopian schemes often find their roots in thorough dissatisfaction with the current situation. Positively, a deep despair about a state of affairs can stem from a sincere attachment to place, and crippling doubt about the path to take can find its origin in a belief of the multiplicity of promising futures. The next chapters unpack these ambiguities, their implications for future strategy.

As usual, affect can be cause and effect, and versions of past, present, and future entangle in ways identifying a community. Stories about past or future can trigger fear and hope, while widespread fear, nostalgia, and high expectations can crystallize stories that proliferate in public discourse (Bangstad et al., 2019; Vanheule, 2016). Unease about the present can lead to anxiety, to fear and anger. Unease can make communities cling to their hopes, find signs of positive evolution. For psychoanalysis, it comes as no surprise that anxiety functions as an almost universal language, a most common translation of unease, non-understanding, non-observation, loss of certainty, of repression, or of unexpected intrusions into the imaginary unity of community life (Burgess, 2017; Vergote & Moyaert, 1988). Collective responses to a variety of external and internal stimuli where thinking and feeling are not certain about categorization lightly translate into anxiety. Anxiety and ambiguity that cannot be easily dispelled might veer into very different directions, each promising clarity of cognition and affect.

Fear of the future is almost certainly overdetermined, infused with a panoply of emotions and fed by a multitude of different narratives. The future is by definition uncertain, and what exists now, or what haunts us from the past, can be projected into the future, while futures can be imagined to extend their influence to the present. What is cherished, can be lost; what is missing, can be gained; and what is unknown, can cast its spell from the future. Futures, and more so collective futures where that collective is under pressure, provoke anxiety (M. Hulme, 2009; Hutchison, 2016). A future unknown can inspire fear, hope, desire, or detachment and disaffection. Many things might be projected simultaneously, some unconsciously; hence, tracing affects back to the future tends to be daunting, and dispelling the notion that nothing in the future is waiting for us or commanding us, tends to be harder than the rationalists present it (Ferreday & Kuntsman, 2011; Žižek, 2009).

Communities have only a few options in facing the future (see Chapter 10). They can look away, or pretend the future is a mere extension of the present, or they can face it and emphasize either hope or fear. A classic modernist reaction is to rationalize and institutionalize the drive for certainty and security (Groarke, 2013; Heath-Kelly, 2019). None of these options are perfect, as in satisfying and stabilizing the community, allaying fears, guaranteeing the preservation of public goods, and minimizing risk. Fetishizing the tools supposed to generate certainty is bound to simplify observation and self-observation, to multiply blind spots, and to reinforce repression and anxiety (McBride, 2007; Scott, 2020). Disavowing the future is not an option, as it is bound to happen. What can be feared is the oscillation itself, between fear and hope, between other affective and cognitive polarities, the instability emanating from it, in public discourse and within governance. The ambivalence (see Chapter 7) can be what must be denied or rationalized later. Without new Master Signifiers, it is not likely to go away, however, and a more productive approach might be to explore those preformed paths that exist in the collective unconscious, and which determine where affective and discursive polarities tend to flip, rather than stabilizing themselves by avoiding confrontation and moving sideways (see Chapter 11).

The affective position of the governance system itself helps determine what it can do in facing and shaping the future. What people feel about their governance system, the role it is expected to play, and the trust bestowed upon it predefines which positions might develop in the system. An example of current relevance is the growing suspicion toward government and its experts, and the drive toward localism, self-organization, and small government, affective strategies to treat government as the enemy, to interpret an existing split between community and governance as unjustifiable (Couperus et al., 2023). One can recognize this easily as a denial of the necessary self-splitting of community and governance, as a desire for an imagined past where the subject was not divided (Gunder & Hillier, 2009).

If governance is the enemy, its role is limited. If it is feared, this, too, represents a limitation of possible roles and possible futures (see Chapter 11). Fear *of* governance coexists with fear *in* governance. This can turn into paranoia, when cohesion is recognized and intentions are ascribed where they do not exist (Žižek, 1992). Mari Ruti (2012), in Lacanian fashion, elegantly exposes how administrative structures can function as Big Other, limiting and structuring our lives in ways suggesting intention even where there is none, where effects come about through myriad rules and interactions. We do things that require an administrative response; we follow rules assuming there is someone scrutinizing our compliance with them; and we want to believe administrative rhetoric suggesting a unified identity and unity of intention. If we shift perspective, and look inside administration, one might find a mirrored paranoia (Clarke, 2005; Samier, 2014). The environment, the world of noisy customers, citizens, and residents can look like it is unified, like it is speaking with one voice, a feared voice blaming those serving the community for everything, expecting contradictory and impossible things. Hiding intention and avoiding interaction are common responses, which densify the mystery of governance and reinforce the role of Big Other, of impersonal systems of control with a strong yet hidden personality.

Desire

Speaking of affect is speaking of desire, and desire itself can be felt as affect. Yet, for psychoanalysis, not all desire is felt, is known, can be understood as something and grasped as an object and orientation of will. What is desired is never identical to what is articulated as desirable in speech, which, for governance is rather relevant (Gunder & Hillier, 2009). Governance, in mainstream political theory, tends to be presented as deliberation of preferences, where people know what they want. Sometimes, the community shares a vision, wants something as a collective, and that can crystallize into a desirable future (McAfee, 2022; Urbinati, 2000).

For us, stories emerge that can create a desiring community, and that community deploys stories and the tools of governance to see what it desires and how to achieve it. Ambivalence (see Chapter 7) never fully retreats, and the desiring community can vanish. We can add to our discussion of community in the previous chapters that story itself is not enough to form a community, that it only emerges when a collective desire co-evolves in those stories. One can also directly infer this from the assertion that what is real must be felt as real, as what is felt is never disconnected from desire. And, following Lacan, the point becomes more obvious if we accept that identity, as Ego and as subject, only comes into being because of the unceasing force of desire, the desire to be, to be with, to become (Valdre, 2019; Van Haute, 2002).

Recognizing the centrality and the dynamics of desire is not always possible in governance, as it would hamper the functioning of governance. Accepting the possibility of unconscious desire, as actor and as collective, renders modernist ideas of statecraft immediately useless. It does not, however, undermine democracy, as people were perfectly fine with the presence of the unconscious, until modernist fictions banished it from reality (Eberle, 2019; Glynos & Stavrakakis, 2010). Governance, moreover, must operate on the basis of translations that try to capture what is desired into what can be understood and actioned upon within the means of governance. Sustainability can function as a Master Signifiers, to guide various policy domains and decisions, but when it comes to the selection and sustainability of one thing and another, it also pushes for a stepwise specification. Worries about the future of the planet then translate into a slightly greener neighborhood, the shade of green being slightly different because of reintroduced native species (Béné et al., 2018; Gunder, 2006).

Lacan compels us to see more dynamism. For him, the *objet petit a* becomes a symbol of desire and its embodiment at the same time. The concept shows up in Lacan's Seminar V, on the unconscious, and becomes a constant in his thinking, notably taking a central role in Seminar XI, on the fundamental concepts of psychoanalysis. The *objet petit a* must exist in the imaginary and symbolic order, in order to function as an attractor of desire, and it is of the order of the Real, as it is an eruption of something that could not find a place in the symbolic and imaginary order (De Kesel, 2009; Gunder, 2005; Hook, 2017; Lacan, 2017[1957]). Sustainability can be embodied by that Swedish project, seen in that magazine one likes, representing a club one would like to be part of, that project led by that designer one would like to be. For the city, the project can reinforce a positive self-image or allay fear about a negative image, and the fact of formulating a higher and newer ideal for the urban environment becomes a signal for an aspirational identity (Gunder, 2004). Later, sustainability as embodied in that project can be captured by administrative experts who make it implementable by translating its features and goals into the concepts and tools available, maybe by defining targets in the quantitative and monetary terms of ecosystem services, maybe by lecturing residents in neighborhood meetings on the value and reality of both the Master Signifier and the translation into this project.

A problem for those strategic thinkers in governance who must manage all this, is that desire tends to move. A second Lacanian complication comes with the likelihood that not the project itself functions as an *objet petit a*, but rather a quality that might not be understood or even consciously perceived by those most ardently embracing it, while the actual reasons for embrace are more varied and ambivalent than what can be allowed into official rhetoric and policy tools related to the project. Similarly, any Master Signifier can lose its capacity to gloss over different interpretations; it can lose its attraction and capacity to produce ever-shifting *objets petit a*.

Concluding

The nature of intentionality in governance cannot be understood without reference to narrative mediating rationality and affect, shaping both. Thinking and organizing in governance allow for collective intentionality, an intentionality that moves through stories, explaining the world and possible futures, and stories speaking of communities that are capable of organizing themselves and their futures. Narratives in governance and the governance system itself can be invested with affect, and they can trigger diverging affects in the community, backlash against them stemming in some cases from affect-shaping arguments. Elsewhere, arguments carry affect and a punch, through associations with other ideas and desires, hopes and fears, histories overt and covert.

Ego, Id, and Superego can appear in governance, and avail themselves of stories, of arguments and emotion. Desires *in* the community and *for* the community have origins where individual and group experiences, as well as the signs and stories available to the community are entwined. Identities framing those desires are themselves the product of contingent discursive evolutions. Affect can spread in a community, in a governance system, for reasons not well understood from within, possibly requiring new forms of reflexivity and second-order observation to trace and disentangle (see Chapter 11).

For us, community is not always there, as Voice and desire are not always there. What appears as real (see previous chapter) becomes so through interaction and coexistence, through sharing places and stories, through the gradual formation of collective desires. In that manner, communities can develop shared perspectives on reality, ways in which they can test each other's reality through knowing and feeling. However, the necessary split between governance and community alienates the community from itself and gives it the power to transform itself toward a desired state. That split simultaneously generates a friction within the self, a barrier for desire to recognize itself completely in governance and in what governance can do. Where collective futures are considered, where fear and hope are thus confronted directly (see Chapter 10), the self-splitting becomes most powerful and most risky, most capable of creating new worlds, and most likely to trip over the Real of its internal division.

Even the most boring routines in governance, seemingly disconnected from any real-life meaning or aspiration, let alone from big visions and desires, can carry an affective investment (Åkerström et al., 2021), and can trigger affective (maybe angry) responses. Seemingly bland realities in the background can be invested with strong feelings and can support identities. All of them, routines and backgrounds (see Chapter 3), can draw attention to themselves, can be reinvested with affect, can become embroiled in heated debates. Their role in governance can create effects in the larger world, which can direct attention to the realities of governance. Affect in the community, new narratives, and new observations of governance can bring dead institutions back to life. Desires unfulfilled or, indeed, fulfilled, circle back to their own assumptions and draw attention to that which escaped observation, discussion, and that which seemed lifeless before. In the next chapter, we investigate the way governance systems manage these dynamics of ossification and revival, through shifting patterns of inclusion and exclusion.

This succinct treatment of inclusion and exclusion, of diversity and lack thereof, in governance and community, sets the stage for our discussion of governance paths and evolutions in Chapter 8, and for our perspective on future strategies in Chapter 10. Meanwhile, our final chapter, on the possibility of community therapy, eschews formulas for participation, diversity, and inclusion as the royal road to resolving community problems and traversing collective fantasies.

Nevertheless, the unity of the perspective on reality at work in governance, and the drawing of external and internal boundaries for the community, are crucial considerations for any attempt to go beyond modernist understandings of policy and administration, and beyond essentializing, and hence essentially conservative, concepts of community. At the most elementary level, this pertains to the definition of insiders and outsiders, and the paths available to switching categories. What does one have to do, to say, to believe, and for how long, to be a member of the community, to have any voice, and what does it take to become an actual insider, someone who might not only be observed but could also come up with new ideas and take initiative? What is the space for outsiders to do their own thing, and when does insider-dominated governance impose limits?

In most places, it does not take decades to be accepted, and formal democratic institutions support political inclusion even where cultural boundaries remain. Mechanisms of inclusion and exclusion remain. Intentional and unintentional, illegal and otherwise, exclusions persist that go against the spirit and the letter of the law, or, against the values of the larger community. Where the pattern of inclusion and exclusion in governance does not seem too objectionable, and not many feel excluded in loathsome ways, this is no guarantee for the future, as the pattern itself does not hold the key to its own transformation. Such a key must be found in the cultivation of learning, of truly diverse perspectives in governance, which can continuously reflect on the appropriateness of the existing inclusions/exclusions and the continued relevance of the implied categories.

References

Åkerström, M., Jacobsson, K., Andersson Cederholm, E., & Wästerfors, D. (2021). *Hidden Attractions of Administration: The Peculiar Appeal of Meetings and Documents*. Taylor & Francis. https://doi.org/10.4324/9781003108436

Allmendinger, P. (2017). *Planning Theory* (3rd ed., p. 346). Macmillan Education UK. https://books.google.ca/books?id=XQAoDwAAQBAJ

Alvesson, M. (1987). Organizations, culture, and ideology. *International Studies of Management & Organization, 17*(3), 4–18. www.tandfonline.com/doi/abs/10.1080/00208825.1987.11656459

Alvesson, M., & Einola, K. (2019). Warning for excessive positivity: Authentic leadership and other traps in leadership studies. *The Leadership Quarterly, 30*(4), 383–395. https://doi.org/10.1016/j.leaqua.2019.04.001

Alvesson, M., & Spicer, A. (2016). *The Stupidity Paradox: The Power and Pitfalls of Functional Stupidity at Work*. Profile Books.

Anderson, B. (2021). Affect and critique: A politics of boredom*. *Environment and Planning D: Society and Space, 39*(2), 197–217. https://doi.org/10.1177/02637758211002998

Bangstad, S., Bertelsen, B. E., & Henkel, H. (2019). The politics of affect: Perspectives on the rise of the far-right and right-wing populism in the West. *Focaal, 2019*(83), 98–113. https://doi.org/10.3167/fcl.2019.830110

Béné, C., Mehta, L., McGranahan, G., Cannon, T., Gupte, J., & Tanner, T. (2018). Resilience as a policy narrative: Potentials and limits in the context of urban planning. *Climate and Development, 10*(2), 116–133. https://doi.org/10.1080/17565529.2017.1301868

Benwell, B. (2006). *Discourse and Identity*. Edinburgh University Press.

Bovens, M. A. P. (1998). *The Quest for Responsibility: Accountability and Citizenship in Complex Organisations*. Cambridge University Press.

Brennan, T. (2004). *The Transmission of Affect*. Cornell University Press.

Brockmeier, J. (2002). Remembering and forgetting: Narrative as cultural memory. *Culture & Psychology, 8*(1), 15–43. https://doi.org/10.1177/1354067X0281002

Burgess, J. P. (2017). For want of not: Lacan's conception of anxiety. In E. Eklundh, A. Zevnik, & E.-P. Guittet (Eds.), *Politics of Anxiety* (pp. 17–36). Rowman & Littlefield.

Clarke, J. (2005). Performing for the public: Doubt, desire, and the evaluation of public services. In P. du Gay (Ed.), *The Values of Bureaucracy* (pp. 211–232). OUP.

Clegg, S., e Cunha, M. P., Munro, I., Rego, A., & de Sousa, M. O. (2016). Kafkaesque power and bureaucracy. *Journal of Political Power, 9*(2), 157–181. https://doi.org/10.1080/2158379X.2016.1191161

Consigny, S. (1974). Rhetoric and its situations. *Philosophy & Rhetoric, 7*(3), 175–186.

Couperus, S., Tortola, P. D., & Rensmann, L. (2023). Memory politics of the far right in Europe. *European Politics and Society, 24*(4), 435–444. https://doi.org/10.1080/23745118.2022.2058757

De Kesel, M. (2009). *Eros and Ethics: Reading Jacques Lacan's Seminar VII*. State University of New York Press.

Drabeck, T. E., & Quarantelli, E. L. (2008). Scapegoats, villains, and disasters. *Crisis Management, 3*, 146–153.

Eberle, J. (2019). Narrative, desire, ontological security, transgression: Fantasy as a factor in international politics. *Journal of International Relations and Development, 22*(1), 243–268. https://doi.org/10.1057/s41268-017-0104-2

Eco, U. (1979). *The Role of the Reader: Explorations in the Semiotics of Texts*. Indiana University Press.

Ferguson, J. (1994). *The Anti-politics Machine: "development," Depolicization, and Bureaucratic Power in Lesotho*. University of Minnesota Press.

Ferreday, D., & Kuntsman, A. (2011). Haunted futurities. *Borderlands*, *10*(2). https://go.gale.com/ps/i.do?p=AONE&sw=w&issn=14470810&v=2.1&it=r&id=GALE%7CA276187004&sid=googleScholar&linkaccess=abs

Fieschi, C., & Heywood, P. (2004). Trust, cynicism and populist anti-politics. *Journal of Political Ideologies*, *9*(3), 289–309. https://doi.org/10.1080/1356931042000263537

Fineman, S. (Ed.). (2000). *Emotion in Organizations* (2nd ed). Sage.

Fink, B. (1999). *A Clinical Introduction to Lacanian Psychoanalysis: Theory and Technique*. Harvard University Press.

Fink, B. (2013). *Against Understanding, Volume 2: Cases and Commentary in a Lacanian Key*. Routledge. https://doi.org/10.4324/9781315884035

Fischer, F. (2009). *Democracy and Expertise: Reorienting Policy Inquiry*. Oxford University Press.

Flyvbjerg, B. (2001). *Making Social Science Matter: Why Social Inquiry Fails and How It Can Succeed Again*. Cambridge University Press.

Flyvbjerg, B. (2004). Phronetic planning research: Theoretical and methodological reflections. *Planning Theory & Practice*, *5*(3), 283–306. https://doi.org/10.1080/1464935042000250195

Freud, S. (1989). *The Ego and the Id* (J. Strachey, Ed.). Norton.

Freud, S. (1990). *Group Psychology and the Analysis of the Ego* (J. Strachey & S. P. of H. P. Gay, Eds.; The Standard ed.). W. W. Norton & Company.

Freud, S. (2003). *The Uncanny* (H. Haughton, Ed.; D. McLintock, Trans.; Illustrated ed.). Penguin Classics.

Freud, S. (2004a). *Mass Psychology*. Penguin UK.

Freud, S. (2004b). *Totem and Taboo* (2nd ed.). Routledge. https://doi.org/10.4324/9780203164709

Freud, S. (2014). *The Neuro-Psychoses of Defence*. Read Books Ltd.

Freud, S. (2015). *Civilization and Its Discontents*. Broadview Press.

Freud, S. (2020). *The Interpretation of Dreams: The Psychology Classic*. John Wiley & Sons.

Freud, S. (2023). *Psychopathology of Everyday Life*. BoD – Books on Demand.

Glynos, J., & Stavrakakis, Y. (2010). Politics and the unconscious – An interview with Ernesto Laclau. *Subjectivity*, *3*(3), 231–244. https://doi.org/10.1057/sub.2010.12

Gould, W., Sherman, T. C., & Ansari, S. (2013). The flux of the matter: Loyalty, corruption and the 'everyday state' in the post-partition government services of India and Pakistan*. *Past & Present*, *219*(1), 237–279. https://doi.org/10.1093/pastj/gts045

Green, A. (1999). *The Fabric of Affect in the Psychoanalytic Discourse*. Routledge. https://doi.org/10.4324/9780203360057

Groarke, S. (2013). *Managed Lives: Psychoanalysis, Inner Security and the Social Order: Psychoanalysis and the Administrative Task*. Routledge. https://doi.org/10.4324/9781315880150

Grube, D. (2014). The gilded cage: Rhetorical path dependency in Australian politics. In J. Uhr & R. Walter (Eds.), *Studies in Australian Political Rhetoric* (pp 99–118).

Gunder, M. (2004). Shaping the planner's ego-ideal: A Lacanian interpretation of planning education. *Journal of Planning Education and Research*, *23*(3), 299–311. https://doi.org/10.1177/0739456X03261284

Gunder, M. (2005). Lacan, planning and urban policy formation. *Urban Policy and Research*, *23*(1), 87–107. https://doi.org/10.1080/08111140042000335287

Gunder, M. (2006). Sustainability: Planning's saving grace or road to perdition? *Journal of Planning Education and Research, 26*(2), 208–221. https://doi.org/10.1177/0739456X06289359

Gunder, M., & Hillier, J. (2009). *Planning in Ten Words or Less: A Lacanian Entanglement with Spatial Planning*. Ashgate Publishing, Ltd.

Heath-Kelly, C. (2019). Forgetting ISIS: Enmity, Drive and Repetition in Security Discourse. In D. B. Monk (Ed.), *Who's Afraid of ISIS?*. Routledge.

Holstein, J. A., & Gubrium, J. F. (1999). *The Self We Live By: Narrative Identity in a Postmodern World*. Oxford University Press.

Hook, D. (2017). *Six Moments in Lacan: Communication and Identification in Psychology and Psychoanalysis*. Routledge.

Hulme, M. (2009). *Why We Disagree about Climate Change: Understanding Controversy, Inaction and Opportunity*. Cambridge University Press.

Hulme, R., & Hulme, M. (2012). Policy learning? Crisis, evidence and reinvention in the making of public policy. *Policy & Politics, 40*(4), 473–489. https://doi.org/10.1332/030557312X645757

Hunter, S. (2015). *Power, Politics and the Emotions: Impossible Governance?*. Routledge-Cavendish.

Hutchison, E. (2016). *Affective Communities in World Politics: Collective Emotions after Trauma*. Cambridge University Press. https://doi.org/10.1017/CBO9781316154670

Jun, J. S. (2012). *The Social Construction of Public Administration: Interpretive and Critical Perspectives*. State University of New York Press.

Knudsen, D. C., Rickly, J. M., & Vidon, E. S. (2016). The fantasy of authenticity: Touring with Lacan. *Annals of Tourism Research, 58*, 33–45. https://doi.org/10.1016/j.annals.2016.02.003

Lacan, J. (2014). *Anxiety: The Seminar of Jacques Lacan, Book X* (J.-A. Miller, Ed.; A. R. Price, Trans.; 1st ed.). Polity.

Lacan, J. (2017[1957]). *The formations of the unconscious* (edited by J. A. Miller). Polity Press.

Latour, B. (2004). *Politics of Nature: How to Bring the Sciences into Democracy*. Harvard University Press. https://doi.org/10.4159/9780674039964

Luhmann, N. (2018). *Organization and Decision*. Cambridge University Press.

Massumi, B. (2015). *Politics of Affect*. John Wiley & Sons.

McAfee, N. (2022). Public Philosophy and Deliberative Practices. In *A Companion to Public Philosophy* (pp. 134–142). John Wiley & Sons, Ltd. https://doi.org/10.1002/9781119635253.ch14

McBride, P. C. (2007). Introduction: The Future's Past—Modernism, Critique, and the Political. In P. C. McBride, R. W. McCormick, & M. Žagar (Eds.), *Legacies of Modernism: Art and Politics in Northern Europe, 1890–1950* (pp. 1–13). Palgrave Macmillan US. https://doi.org/10.1057/9780230603189_1

McCormick, J. P. (2011). *Machiavellian Democracy*. Cambridge University Press.

McGowan, T. (2022). Sublimating the commodity. *Distinktion: Journal of Social Theory, 23*(2–3), 359–377. https://doi.org/10.1080/1600910X.2022.2054447

Merkley, E. (2020). Anti-intellectualism, populism, and motivated resistance to expert consensus. *Public Opinion Quarterly, 84*(1), 24–48. https://doi.org/10.1093/poq/nfz053

Mesquita, B., Boiger, M., & De Leersnyder, J. (2017). Doing emotions: The role of culture in everyday emotions. *European Review of Social Psychology, 28*(1), 95–133. https://doi.org/10.1080/10463283.2017.1329107

Neuman, W. R. (2007). *The Affect Effect: Dynamics of Emotion in Political Thinking and Behavior*. University of Chicago Press.

Newman, J. E., & Clarke, J. H. (2009). *Publics, Politics and Power: Remaking the Public in Public Services*. Sage.

Patz, R., Thorvaldsdottir, S., & Goetz, K. H. (2022). Accountability and affective styles in administrative reporting: The Case of UNRWA, 1951–2020. *Journal of Public Administration Research and Theory, 32*(1), 111–129. https://doi.org/10.1093/jopart/muab024

Rancière, J. (2023). Politics, Identification and Subjectivization. In M. J. Liger (Ed.), *Identity Trumps Socialism*. Routledge.

Roelvink, G. (2010). Collective action and the politics of affect. *Emotion, Space and Society, 3*(2), 111–118. https://doi.org/10.1016/j.emospa.2009.10.004

Ruti, M. (2012). *The Singularity of Being: Lacan and the Immortal Within*. Fordham University Press.

Salvatore, S., Picione, R. D. L., Bochicchio, V., Mannino, G., Langher, V., Pergola, F., Velotti, P., & Venuleo, C. (2021). The affectivization of the public sphere: The contribution of psychoanalysis in understanding and counteracting the current crisis scenarios. *International Journal of Psychoanalysis and Education: Subject, Action & Society, 1*(1), Article 1. https://doi.org/10.32111/SAS.2021.1.1.2

Samier, E. A. (2014). *Secrecy and Tradecraft in Educational Administration: The Covert Side of Educational Life*. Routledge. https://doi.org/10.4324/9780203387085

Scott, J. C. (2020). *Seeing Like a State: How Certain Schemes to Improve the Human Condition Have Failed*. Yale University Press.

Soler, C. (2015). *Lacanian Affects: The Function of Affect in Lacan's Work*. Routledge. https://doi.org/10.4324/9781315731797

Tanggaard, L. (2013). The sociomateriality of creativity in everyday life. *Culture & Psychology, 19*(1), 20–32. https://doi.org/10.1177/1354067X12464987

Tewdwr-Jones, M., & Allmendinger, P. (1998). Deconstructing communicative rationality: A critique of Habermasian collaborative planning. *Environment and Planning A: Economy and Space, 30*(11), 1975–1989. https://doi.org/10.1068/a301975

Urbinati, N. (2000). Representation as advocacy: A study of democratic deliberation. *Political Theory, 28*(6), 758–786. https://doi.org/10.1177/0090591700028006003

Valdre, R. (2019). *On Sublimation: A Path to the Destiny of Desire, Theory, and Treatment*. Routledge. https://doi.org/10.4324/9780429478048

Van Assche, K., Beunen, R., & Gruezmacher, M. (2024). *Strategy for Sustainability Transitions: Governance, Community and Environment*. Edward Elgar Publishing.

Van Assche, K., Gruezmacher, M., Marais, L., & Perez-Sindin, X. (2023). *Resource Communities: Past Legacies and Future Pathways*. Taylor & Francis. https://books.google.ca/books?id=4EHUEAAAQBAJ

Van Haute, P. (2002). *Against Adaptation: Lacan's "Subversion" of the Subject*. Other Press.

Vanheule, S. (2016). Capitalist discourse, subjectivity and Lacanian psychoanalysis. *Frontiers in Psychology, 7*. https://doi.org/10.3389/fpsyg.2016.01948

Vanheule, S., & Verhaeghe, P. (2009). Identity through a psychoanalytic looking glass. *Theory & Psychology, 19*(3), 391–411. https://doi.org/10.1177/0959354309104160

Vergote, A., & Moyaert, P. (1988). *Psychoanalyse: De mens en zijn lotgevallen*. DNB/ Pelckmans.

Viroli, M. (1998). *Machiavelli*. OUP.

Wagenaar, H. (2015). 22 Transforming perspectives: The critical functions of inter-pretive policy analysis. In F. Fischer, D. Torgerson, A. Durnová, & M. Orsini (Eds.), *Handbook of Critical Policy Studies* (pp 422–440). Edward Elgar.

Yanow, D. (2000). *Conducting Interpretive Policy Analysis* (Vol. 47). Sage.

Zepf, S. (2011). About rationalization and intellectualization. *International Forum of Psychoanalysis, 20*(3), 148–158. https://doi.org/10.1080/0803706X.2010.550316

Žižek, S. (1992). *Looking Awry: An Introduction to Jacques Lacan through Popular Culture*. MIT Press.

Žižek, S. (2002). *Did Somebody Say Totalitarianism?: Five Interventions in the (mis) use of a Notion*. Verso.

Žižek, S. (2009). *In Defense of Lost Causes*. Verso. http://public.ebookcentral.proqu est.com/choice/publicfullrecord.aspx?p=5176960

Žižek, S. (2011). *Living in the End Times*. Verso Books.

6   Inclusion, exclusion, and diversity

Communities cannot be entirely inclusive, as communities require structures and boundaries. They cannot be fully inclusive as decisions about what could be included are still decisions. Decisions are by definition jumps into the unknown (Luhmann, 2018). Reasoning alone cannot produce a decision, that is, interpretations of the pros and cons of recognizable alternatives. Those alternatives must be delineated, which requires, again, an arbitrary reduction of contingency, a choice. What can be deduced, is never a decision (Rasche, 2008). This insight from systems theory is relevant for the understanding of governance, as it undercuts the possibility of rational decision-making, while psychoanalysis modified the image of the rational decision-maker long before.

The need to make decisions in governance means that the before and after remain only loosely connected, while ideas and interests that have to be represented and translated into policy come out disfigured. Moreover, governance becomes a system in itself, with actors responding to each other's realities and creating a reality of governance, where simplified versions of the world become amenable to decision-making, where desires can be translated into actionable items on an agenda (Brans & Rossbach, 1997; Valentinov, 2014). Stories in the system and realities as experienced outside, in the community, will differ. In addition, what is needed to balance interests, and to make something happen that is demanded by voters, is not always knowable by those voters (Mansfield, 1998) (and see earlier chapters).

Even if, in governance, the community can represent itself to itself, and strategize toward a better self, there will always be distance between the initial image of self and the one produced and reflected through governance (Luhmann, 1987; Rasche & Seidl, 2020). What comes out of governance in terms of decisions and in terms of implementation and action, is never what the community wanted, and the perceived pattern of inclusion and exclusion is never what it was desired to be (Lyons et al., 2019). In this process, communities can disrupt their imaginary unity, through governance that was perhaps intended to do the opposite. In democratic systems, many different voices are supposed to disrupt an imaginary unity that is simultaneously desired

DOI: 10.4324/9781032696676-6

by the community. This tension can enrich learning and discussion, and it can become destabilizing if communities are not willing to face the impossibility of perfect unity and harmony (Gentile, 2020; Gunder & Hillier, 2009). Each imperfect pattern of inclusion and exclusion, shaped by a history of choices in policy and institutional design, is thus bound to trigger responses and carries the seeds of its own destruction (Fleisher, 1995).

If representation is *felt* as too far from perfect, this can aggravate tensions in the community, and lead to outright conflict. Histories of conflict, in turn, can render every decision on a legitimate pattern of inclusion sensitive (Frosh, 2013). Polarization can push toward simplification of identities and inclusion patterns, rendering practical collaboration difficult. Affect accompanies any decision and discussion on inclusion and exclusion, as it directly concerns the representation of values and identities, and as the overall pattern is expected to represent community identity. The unity of voice that is expected by some, can be felt as imposed by others; the diversity appreciated by some can be felt as non-community by others (Hoggett, 1997; McGowan, 2023).

The nature of decisions as non-reducible to antecedents and the fact of co-evolution in governance make for highly contingent patterns of inclusion, resulting from decisions *on* inclusion but more so as effects of decisions on other topics. Psychoanalysis would add that officially dismissed options, formally excluded voices, and supposedly forgotten histories are rarely completely forgotten, excluded, and dismissed. They can linger on, in memory, in the collective unconscious, in the community and, in smaller pockets, in governance. Narratives not accepted or deemed irrelevant in governance can be affectively invested by those feeling alienated by that system. The more disaffected one feels, the more intensely one can be attracted by all that was ignored and dismissed by a Big Other one does not respect (Hoggett, 2013).

Similarly, affects can linger, either attached to narratives or not. They can transfer to different topics and stories. Unexamined affect can intensify in the margins, and it can remain present in the ambivalence of a majority (see next chapter). It can hold the potential for creating a new majority, or a new mode of expression and mobilization. Arguments and affects pertaining to exclusion can precede a call for inclusion, and they can crystallize with the formation of new subjectivities around a surviving and spreading affect. What can be disavowed for a long time, what seemed unthinkable, can appear as self-evident, translatable into viable policy (Kinnvall & Svensson, 2022). The enjoyment barred before can be an enjoyment of exclusion, which can be felt as an enjoyment of purity, or of natural values, differences, and hierarchies. Classism, sexism, and racism always found ways to naturalize themselves (McAfee, 2021; McKenzie, 2020). What was disavowed earlier can transform into something desirable, or its normalization can open the door to very different futures. It does not have to be pursued directly, perhaps in the vein of a return to an imagined past; it can produce a variety of new futures by lifting the veil of repression.

The flip side is that what was cherished before, can now look burdensome. Regulatory systems built to prevent past mistakes or to protect shared assets can now feel like unreasonable constraints on the enjoyment of freedom or exclusion. Checks and balances, democratic procedures, and human rights can feel irrelevant compared to the promised joy of purity, unity, freedom, or whatever Master Signifier becomes central to the awoken affective community (Roelvink, 2010; Urbinati, 2014). These can be felt as an impediment, as an injustice imposed on the real community, even if that community only started to recognize itself recently, and even if those feeling the pain now largely overlap with those inflicting it.

Memory and identity revisited

Policy alternatives look and feel different as time passes, in a new context. New options appear, while the governance system and the community taking the decision change. Psychoanalysis knows that no decision is forever, and that what is rejected tends to survive in the shadows. In governance, those shadows do not have to be deep, as memory functions exist, as do public discourse, and arenas where opposing discourses can explore alternative memories and hidden affect.

Discursive dynamics in the community make forever shifting identities present as actors in governance, hence a different form of remembering and forgetting of the past. What looks like inclusion can feel like exclusion later; what appeared earlier as a reasonable decision might look like an insidious move by a dominant group (Williams, 1985). Groups can reinterpret their position as oppressed after decades of feeling themselves part of the community; conversely, they can reinterpret a position as victim in a more complex manner later. Meanwhile, consciously veiled forms of marginalization can be revealed later, through shifting positions of power and advocacy, but also by the continuous restructuring of memory already taking place in governance (Sanchez & Moore, 2015). Thus, perspectives shift and identities transform, yet this does not prevent communities from discovering that injustices did occur, that patterns of inclusion and exclusion did harm people, did trample the espoused principles of justice and the fair distribution of benefits and public goods.

Lacanian theory underlines the revealing and obscuring functions of Master Signifiers such as unity, purity, freedom, and property, but also inclusion, social justice, sustainability, and democracy itself (Bailly, 2012; Hook, 2017). The certainty desired with regard to correct forms of inclusion is never real, never stable, and this productive fiction is stabilized by other productive fictions, including notions of the knowability of governance and community, of self-evident ideas of justice and fair distribution, of inclusive governance (Greene, 2013). In the case of inclusivity itself, it is rather obvious that not everything and not everyone can be included, can be active, and can be represented. This basic understanding is compatible with most standard narratives of democracy and its history, and with the empirical observation of early state development (Scott, 2017).

An individual can be represented in certain aspects of her identity, certain values and desires, fears and hopes. Structures of policy domains, topics, temporalities, patterns of expertise, of participation and representation, traditions of public discourse and debate, and levels and forms of education all contribute to what counts as a citizen, as a group, as a voice, and to how these can connect to processes of collective decision-making, and their transformation (King & Thornhill, 2003).

Direct participation, in contrast to indirect representation, does not alter this reality (Urbinati, 2006). The existence of groups, of large organizations representing them, of leadership representing the membership, reinserts the same dilemmas of distance, reinterpretation, selective understanding, and representation associated with more centralized governance systems based on representation. Calls for participation can again be interpreted as signs that the existing patterns of inclusion are not felt as fair and functional anymore, and that the existing balance of representation and participation is felt to be inadequate (Hoggett et al., 2013). The fiction of perfect inclusion underpins the call.

Power/knowledge and inclusion/exclusion

Following how Michel Foucault sees power and knowledge, as mutually constitutive with governance both codifying and transforming power/knowledge configurations, one can readily see that governance can exert power to change things, to understand, and to make understandings (Foucault, 2002, 2012). Creating worlds, discursively and materially, is within the purview of governance, so the issue of inclusion is not just a matter of representation in governance, but also in the worlds created through governance (Hunter, 2015; Janes, 2016).

Governance instills the power to include and exclude from reality and from decision-making over reality. Master narratives driving governance can directly define what and who should be included, and what should even be recognized as existing, yet this defining power can also lie with other narratives, which could still be supported by the master narrative (Van Assche et al., 2017). The process of decision-making and the process of discursive creation in governance are both processes of exclusion (of alternatives, of realities) that are constitutive of governance, and of bestowing power upon those realities and actors which are included (Hillier, 2002; Jessop, 2014).

If we grasp power as necessary to make governance work, and discourse as never disconnected from power effects, we can recognize the attraction of governance to those interested in power, in and through governance. If we furthermore see, as in the previous chapter, discourse and desire as inseparable, and identities as narrative constructs imbued with desire, a series of connections become visible that have been explored in political theory inspired by psychoanalysis. Desire for knowledge, for finding one's way in the world, for community and belonging, for stability, or for unity, purity, or superiority will find its way in the discursive production in and of governance, its choice of institutions, and its articulation of policy priorities (Glynos, 2010; Gunder & Hillier, 2009; McGowan, 2012; Stavrakakis, 2002).

Recognition of ideas, people, issues, groups, and perspectives that *could* be included in governance, or *could* be affected by means of governance, stems from this knot of power, knowledge, and desire, mediated by ideology and identity (Žižek, 1992, 2006). This knot is not a random entwining, as what is accepted as desire in governance, what is recognized as legitimate discourse and legitimate distribution of discourse and power (hence inclusion and exclusion), will follow from the unique history of the relation between governance and community (McSwite, 1997; Peters & Pierre, 2006). What comes first under which conditions will be determined by that contingent history: Where can desire be bluntly articulated as relevant? Where should an understanding of reality be presented as a reason for action? Where might a reshuffling of power be a legitimate topic of discussion? Inclusion/exclusion is reshuffled in each of these situations, and each starting point will have ramifications for the other realms – power, knowledge, desire (Eberle, 2019; Fortier, 2013).

One can distinguish between deliberate, or intentional, exclusions and inclusions, and in contrast, unintended exclusions. Some might be unobservable. One can further distinguish between legitimate and illegitimate exclusions, where on the illegitimate side one can find not only those produced by knowingly breaking the law, but also those that stem from blind spots and incoherent policies or contradictory values and narratives (see next chapter). Governance can more easily deliberate and modify what was visible and intentional, and it can more readily punish what was illegitimate and intentional (Valentinov et al., 2019; Van Assche, 2013). Second-order observation in governance and in public discourse, in engaged academic reflections, can expose exclusions not visible before, or can expose incoherence in governance, producing problematic exclusions (Fuchs, 2001).

Both post-structuralism and psychoanalysis would indicate that no typology or category is ahistorical, and that no identity can be essentialized to one category in one typology (Newman, 2005; Žižek, 2006). Traumatic histories, a painful blindness of a majority for the suffering of a minority, for the persistent marginalization of a religious or ethnic group, call for action. Policies must be enacted in order to redress historical injustices. Old identities do not survive intact, however. Restoration is not an option where power, knowledge, and desire are entwined, and where co-evolution rules. What will emerge is new, and the newly empowered category and typology will unavoidably produce new blind spots and resistances (Foucault, 2003; Žižek, 2008).

The value of diversity

A different argument for diversity still stands. Pressures to simplify, to purify, and to unify will persist. A unified world, with clear hierarchies, a simple understanding of problems, and desirable solutions to common goods and threats is attractive for decision-makers, administrators, and for a community hoping for stability and identity (Panizza & Stavrakakis, 2020). We want to know and feel we know how things work – how they should work – and what is not clear and simple easily feels and sounds like it is not true, as a potentially dangerous distraction from what *obviously* ought to be the focus of attention (Bangstad et al., 2019). Complexity distracts, diversity slows down, reflexivity confuses, discussion is the gateway to conflict, and conflict is a matter of either bad intentions or stupidity (Merkley, 2020).

Second-order observation, enshrined in governance and fostered in the community, is of the essence in generating counterpressure. Second-order observation comes most easily when a diversity in perspectives is allowed to exist in governance, when reflection and discussion are cultivated (Jessop, 2002) (see also Chapter 4). First-order observation of diverse kinds can remain diverse more easily in the presence of second-order observation, and where stories about efficiency, objectivity, and universality are critically scrutinized (Fuchs, 2001; Hood & Peters, 2004). Any categorization that is presented as immune to historical and cultural forces, and as immune to reflexivity and critical scrutiny, ought to be looked at with appropriate suspicion. Otherwise, the pressure to close the worlds of governance builds, and, with that, the power of governance to restructure society becomes unquestioned, while society is observed from very few angles (Mazzoni, 2023). Other forms of identification, unrecognized or emerging desires, new lifestyles and values, but also brewing resentment and backlash against inclusive policies are at risk of not being observed or interpreted only through the categories dominant at that particular moment (Van Assche et al., 2024). This in turn increases the chances of backlash and of throwing out the baby with the bathwater.

Dialectical learning (see Chapter 4) ought to be defended, and this is more likely to succeed by maintaining a diversity of modes of observation, in governance and of governance, leading to a diversity of truly different perspectives, which can then engage and, ideally, learn (Savski, 2020; Sullivan & McCarthy, 2008; Van Assche et al., 2021). The maintenance of diverse arenas for social learning and the diverse forms of learning distinguished in the previous chapters (comparative, experimental, reflexive, expert, dialectic) can reduce chances of a backsliding into monolithic truths that are bound to rub up against what is truly alive in the community (Henke, 2001). Robust learning infrastructures and diverse learning modes do not only offer a polyphonic mode of reality testing; they also represent conditions for and observation of diversity in perspectives in governance (Bohman, 2006; Luhmann, 1990). The realities constructed in governance become richer, more akin to what is felt in the community. They are also more likely to be different, as a multiplicity of learning modes tends to multiply and amplify difference. Which categories become most relevant for decision-making, as representative of true diversity, of most important distinctions, is not an input for reality testing in governance, or directly for decision-making. Rather, it is a result.

Morality is commonly used by groups, by discourses, to auto-immunize, to evade questioning. Claiming moral superiority can be heartfelt, or it can be more strategic in nature, but the result is the same: an undermining of critical discussion, a degradation of reality testing, an erosion of learning processes in governance (Verschraegen & Madsen, 2013). Hence, a multiplication of blind spots, for narrative and affective trends that, for the sake of democracy, and for the sake of a more broadly understood morality, had better be observed. If moral judgment is tied to one version of diversity, to a preference for one typology, and if this leads to an entrenchment of identity politics, then polarization is around the corner, and finding common ground, or embracing new notions of diversity becomes gradually harder (Bennett & Shapiro, 2002).

References

Bailly, L. (2012). *Lacan: A Beginner's Guide*. Simon and Schuster.

Bangstad, S., Bertelsen, B. E., & Henkel, H. (2019). The politics of affect: Perspectives on the rise of the far-right and right-wing populism in the West. *Focaal, 2019*(83), 98–113. https://doi.org/10.3167/fcl.2019.830110

Bennett, J., & Shapiro, M. J. (Eds.). (2002). *The Politics of Moralizing* (1st edition). Routledge.

Bohman, J. (2006). Deliberative democracy and the epistemic benefits of diversity. *Episteme, 3*(3), 175–191. https://doi.org/10.3366/epi.2006.3.3.175

Brans, M., & Rossbach, S. (1997). The autopoiesis of administrative systems: Niklas Luhmann on public administration and public policy. *Public Administration, 75*(3), 417–439.

Eberle, J. (2019). Narrative, desire, ontological security, transgression: Fantasy as a factor in international politics. *Journal of International Relations and Development, 22*(1), 243–268. https://doi.org/10.1057/s41268-017-0104-2

Fleisher, M. (1995). The ways of Machiavelli and the ways of politics. *History of Political Thought, 16*(3), 330–355.

Fortier, A. -M. (2013). What's the big deal? Naturalisation and the politics of desire. *Citizenship Studies, 17*(6–7), 697–711. https://doi.org/10.1080/13621 025.2013.780761

Foucault, M. (2002). *Archaeology of Knowledge* (2nd ed.). Routledge. https://doi.org/ 10.4324/9780203604168

Foucault, M. (2003). *Society must be Defended: Lectures at the College de France, 1975-76*. Allen Lane The Penguin Press.

Foucault, M. (2012). *Discipline and Punish: The Birth of the Prison*. Knopf Doubleday Publishing Group.

Frosh, S. (2013). Psychoanalysis, colonialism, racism. *Journal of Theoretical and Philosophical Psychology, 33*(3), 141–154. https://doi.org/10.1037/a0033398

Fuchs, S. (2001). *Against Essentialism: A Theory of Culture and Society*. Harvard University Press.

Gentile, J. (2020). Time may change us: The strange temporalities, novel paradoxes, and democratic imaginaries of a pandemic. *Journal of the American Psychoanalytic Association, 68*(4), 649–669. https://doi.org/10.1177/0003065120955120

Glynos, J. (2010). *?Lacan at Work?* (C. Cederstrom & C. Hoedemaeker, Eds.; pp. 13–58). MayFly Books. https://repository.essex.ac.uk/4085/

Greene, M. (2013). A critique of social justice as an archival imperative: What is it we're doing that's all that important? *The American Archivist, 76*(2), 302–334. https://doi.org/10.17723/aarc.76.2.147441214663kw43

Gunder, M., & Hillier, J. (2009). *Planning in Ten Words or Less: A Lacanian Entanglement with Spatial Planning*. Ashgate Publishing, Ltd.

Henke, H. (2001). Freedom ossified: Political culture and the public use of history in Jamaica 1. *Identities, 8*(3), 413–440. https://doi.org/10.1080/1070289X.2001.9962699

Hillier, J. (2002). *Shadows of Power: An Allegory of Prudence in Land-Use Planning*. Routledge.

Hood, C., & Peters, G. (2004). The middle aging of new public management: into the age of paradox?. *Journal of Public Administration Research and Theory, 14*(3), 267–282.

Hoggett, P. (Ed.) (1997). Contested Communities (pp. 3–16). Policy Press. https://bristoluniversitypressdigital.com/edcollchap/book/9781447366645/ch001.xml

Hoggett, P. (2013). Governance and social anxieties. *Organisational and Social Dynamics, 13*(1), 69–78.

Hoggett, P., Wilkinson, H., & Beedell, P. (2013). Fairness and the politics of resentment. *Journal of Social Policy, 42*(3), 567–585. https://doi.org/10.1017/S0047279413000056

Hook, D. (2017). *Six Moments in Lacan: Communication and Identification in Psychology and Psychoanalysis*. Routledge.

Hunter, S. (2015). *Power, Politics and the Emotions: Impossible Governance?* Routledge-Cavendish. https://doi.org/10.4324/9780203798041

Janes, J. E. (2016). Democratic encounters? Epistemic privilege, power, and community-based participatory action research. *Action Research, 14*(1), 72–87. https://doi.org/10.1177/1476750315579129

Jessop, B. (2002). Governance and Meta-governance in the Face of Complexity: On the Roles of Requisite Variety, Reflexive Observation, and Romantic Irony in Participatory Governance. In H. Heinelt, P. Getimis, G. Kafkalas, R. Smith, & E. Swyngedouw (Eds.), *Participatory Governance in Multi-Level Context: Concepts and Experience* (pp. 33–58). VS Verlag für Sozialwissenschaften. https://doi.org/10.1007/978-3-663-11005-7_2

Jessop, B. (2014). From Micro-Powers to Governmentality: Foucault's Work on Statehood, State Formation, Statecraft and State Power. In D Owen (Ed.), *Michel Foucault*. Routledge.

King, M., & Thornhill, E. (2003). *Niklas Luhmann's Theory of Politics and Law*. Palgrave Macmillan.

Kinnvall, C., & Svensson, T. (2022). Exploring the populist 'mind': Anxiety, fantasy, and everyday populism. *British Journal of Politics and International Relations, 24*(3), 526–542. https://doi.org/10.1177/13691481221075925

Luhmann, N. (1987). The representation of society within society. *Current Sociology, 35*(2), 101–108. https://doi.org/10.1177/001139287035002010

Luhmann, N. (1990). *Essays on Self-Reference* (p. 245 Pages). Columbia University Press.

Luhmann, N. (2018). *Organization and Decision*. Cambridge University Press.

Lyons, W. E., Lowery, D., & DeHoog, R. H. (2019). *The Politics of Dissatisfaction: Citizens, Services and Urban Institutions*. Routledge. https://doi.org/10.4324/978131 5288734

Mansfield, H. C. (1998). *Machiavelli's Virtue*. University of Chicago Press.

McAfee, N. (2021). Beyond "Populism": The psychodynamics of antipolitical popular movements. *Populism, 4*(2), 172–198. https://doi.org/10.1163/25888072-bja10020

McGowan, T. (2012). *The End of Dissatisfaction?: Jacques Lacan and the Emerging Society of Enjoyment*. State University of New York Press.

McGowan, T. (2023). Sublimating the Commodity. In H. Richter (Ed.), *Viral Critique*. Routledge.

McKenzie, M. (2020). Affect Theory and Policy Mobility: Challenges and Possibilities for Critical Policy Research. In B. Lingard (Ed.), *Globalisation and Education*. Routledge.

McSwite, O. C. (1997). Jacques Lacan and the theory of the human subject: How psychoanalysis can help public administration. *American Behavioral Scientist, 41*(1), 43–63. https://doi.org/10.1177/0002764297041001005

Merkley, E. (2020). Anti-intellectualism, populism, and motivated resistance to expert consensus. *Public Opinion Quarterly, 84*(1), 24–48. https://doi.org/10.1093/poq/nfz053

Newman, S. (2005). *Power and Politics in Poststructuralist Thought: New Theories of the Political*. Routledge. https://doi.org/10.4324/9780203015490

Panizza, F., & Stavrakakis, Y. (2020). Populism, Hegemony, and the Political Construction of "The People": A Discursive Approach. In P. Ostiguy, F. Panizza, & B. Moffitt (Eds.), *Populism in Global Perspective*. Routledge.

Peters, B. G., & Pierre, J. (2006). Governance, Accountability and Democratic Legitimacy. In A. Benz & I. Papadopoulos (Eds.), *Governance and Democracy*. Routledge.

Rasche, A. (Ed.). (2008). Strategic Realities—The Role of Paradox. In *The Paradoxical Foundation of Strategic Management* (pp. 179–191). Physica-Verlag HD. https://doi.org/10.1007/978-3-7908-1976-2_5

Rasche, A., & Seidl, D. (2020). A Luhmannian perspective on strategy: Strategy as paradox and meta-communication. *Critical Perspectives on Accounting, 73*, 101984–101984. https://doi.org/10.1016/j.cpa.2017.03.004

Roelvink, G. (2010). Collective action and the politics of affect. *Emotion, Space and Society, 3*(2), 111–118. https://doi.org/10.1016/j.emospa.2009.10.004

Sanchez, J. C., & Moore, K. R. (2015). Reappropriating public memory: Racism, resistance and erasure of the Confederate Defenders of Charleston Monument. *Present Tense, 5*(2), 1–8.

Savski, K. (2020). Polyphony and polarization in public discourses: Hegemony and dissent in a Slovene policy debate. *Critical Discourse Studies, 17*(4), 377–393. https://doi.org/10.1080/17405904.2019.1609537

Scott, J. C. (2017). *Against the Grain: A Deep History of the Earliest States*. Yale University Press.

Stavrakakis, Y. (2002). *Lacan and the Political*. Routledge.

Sullivan, P., & McCarthy, J. (2008). Managing the polyphonic sounds of organizational truths. *Organization Studies, 29*(4), 525–541. https://doi.org/10.1177/017084060 8088702

Urbinati, N. (2006). *Representative Democracy: Principles and Genealogy*. University of Chicago Press.

Urbinati, N. (2014). *Democracy Disfigured: Opinion, Truth, and the People*. Harvard University Press. https://doi.org/10.4159/harvard.9780674726383

Valentinov, V. (2014). The complexity–sustainability trade-off in Niklas Luhmann's social systems theory. *Systems Research and Behavioral Science, 31*(1), 14–22.

Valentinov, V., Verschraegen, G., & Van Assche, K. (2019). The limits of transparency: A systems theory view. *Systems Research and Behavioral Science, 36*(3), 289–300. https://doi.org/10.1002/sres.2591

Van Assche, K. (2013). Visible and invisible informalities and institutional transformation. Lessons from transition countries: Georgia, Romania, Uzbekistan. In N. Hayoz & C. Giordano (Eds.), *Informality and Post-Socialist Transition*. Peter Lang.

Van Assche, K., Beunen, R., Duineveld, M., & Gruezmacher, M. (2017). Power/knowledge and natural resource management: Foucaultian foundations in the analysis of adaptive governance. *Journal of Environmental Policy & Planning, 19*, 308–322.

Van Assche, K., Beunen, M., & Gruezmacher, M. (2024). Strategy for sustainability transitions. Edward Elgar.

Van Assche, K., Verschraegen, G., & Gruezmacher, M. (2021). Strategy for collectives and common goods: Coordinating strategy, long-term perspectives and policy domains in governance. *Futures, 128*, 102716–102716. https://doi.org/10.1016/j.futures.2021.102716

Verschraegen, G., & Madsen, M. R. (2013). Differentiation and inclusion: A neglected sociological approach to fundamental rights. *Making Human Rights Intelligible. Towards a Sociology of Human Rights*, 61–80.

Williams, J. (1985). Redefining institutional racism. *Ethnic and Racial Studies*. www.tandfonline.com/doi/abs/10.1080/01419870.1985.9993490

Žižek, S. (1992). *Looking Awry: An Introduction to Jacques Lacan through Popular Culture*. MIT Press.

Žižek, S. (2006). *Interrogating the Real*. Bloomsbury Publishing.

Žižek, S. (2008). *Violence*. Picador.

7   Incoherence and ambivalence

As perfect and perfectly stable patterns of inclusion in governance are productive fictions, so are the aspirations to design governance systems that can eliminate conflict and establish lasting unity and harmony, internally and with the environment ("sustainability"). Indeed, communities and individuals teem with internal contradictions, aspirations, and demands. Feelings about a decision, a policy, about a desired future, are never entirely stable, and even with the same actor, are marked by ambivalence. As we know, governance and community cannot coincide, and a distance between them must be established, making governance an operation based on images and interpretations. Governance systems internalize interpretations of society, and society wants to recognize itself in the governance system, but never fully succeeds. Neither participation nor representation resolves this issue of the divided self in a community. Communities must split themselves into parts that cannot know each other and will exist in friction, in order to function, in order, in fact, to come into existence.

Governance is, moreover, incomplete, as not everything can be reflected in governance, not everything can be understood, organized, and prioritized, and it is impossible to find win-win solutions in every situation. It is incomplete because its powers of understanding, representation, and organization are limited, as are its resources and its capacity to resolve conflicts and to grasp the landscape of identities and desires in the community (Van Assche, Gruezmacher et al., 2022). Finding a stable formula for representation, resolving the issue of diversity and inclusivity forever, would require a consensus and a discursive stability that simply cannot exist (Keating et al., 2023). Incompleteness in this sense generates frictions, as representation is always contested, always subject to revision, and shifting identities and latent affect are easy fodder for political entrepreneurs willing to alter patterns of inclusion for their own or factional benefit (Fieschi & Heywood, 2004).

Governance is not only incomplete, but is also, necessarily incoherent. There are limits to what can be governed, and limits to the processing of complexity in governance (Luhmann, 2018; Luhmann et al., 2004). Dividing, specializing, and compartmentalizing are tricks of the trade that can make

DOI: 10.4324/9781032696676-7

a topic manageable, yet they come at a cost. Coordination becomes harder to achieve, and incoherence is more likely to develop (Brans & Rossbach, 1997). Incoherence can exist for various reasons, technical and otherwise, which will be unpacked in this chapter. Incoherence, it will be argued, can be intentional or not. It is not always a by-product of conflict, complexity, or ignorance. It can be the result of stepwise processes of policy formulation and implementation, of combined application of policies, where what happens in each step is not always to be foreseen by the initial decision-makers. The effects of one policy or one policy domain are never fully observed, and at the same time, typically, the different responsible actors in governance do not observe each other. The reasons for a particular policy and the effect of that policy are similarly varied and similarly unobserved.

Ambivalence, meanwhile, rears its head everywhere in governance. It marks individuals, organizations, governance systems, and communities, observing themselves, trying to identify their own feelings, intentions, and arguments, and those of others. The difficulty, and opportunity, is, once again, the undecidability typical of governance. Decisions need to be taken, something has to be done – but what? Governance exists because at some point a system evolved for collective decision-making. The system survived, and was perceived as legitimate and somehow functional. Yet, who tells the decision-makers how and what to decide? One standard solution in the worlds of bureaucratic and legalist governance is to shift the what to the how, to rely on procedures that are then supposed to suggest the best possible decision, the best course of action. Procedural governance, as well as expert-driven governance, is attractive as it relieves many of the difficulties in deciding (Howlett, 2000; Miller, 2002). Signing is easy.

This conceptual move does not solve the problem of undecidability, however. Decisions, as we know, are jumps. They never follow out of knowledge or logic. A choice remains. That choice cannot be fully rational, as this would suggest there is no choice (Seidl & Becker, 2006). It could still be rational if we assume a stable and unambiguous preference that can then be pursued in a rational manner. This in turn assumes, at the community level, a perfect consensus, not prone to doubt and affective or cognitive oscillations (Green & Shapiro, 1994). Yet, psychoanalysis has taught us that stable preferences, perfect consensus, and perfect self-knowledge regarding preferences are very rare occurrences. Ambivalence is prevalent, and reinterpretation of decisions, stories, and identities is always at play, through social and cultural processes, collective experiences, and through the functioning of governance itself (Van Assche et al., 2011). What seems fair for a majority can feel very different after new stories circulate, after a group of people starts to identify even slightly differently. It can also appear different because incoherence and complexity in governance produce outcomes that do not meet the expectations of the majority, leading to a rejection, not of that particular outcome but of the underlying story.

In this chapter, we take some time to dissect the relation between ambivalence and incoherence in governance, and how it affects stability and instability, harmony and conflict in governance. Productive fictions will appear as often productive, and we will bring in once again, late systems theorist Niklas Luhmann, and his concept of de-paradoxification, or the unfolding of paradox.

Incoherence

Complex governance systems, working in and for complex societies, always have a problem managing complexity (Duit & Galaz, 2008). In order to understand and organize societies, they create internal complexity, through the multiplication of knowledges, concepts, departments, procedures, and roles that then, of course, require coordination (Luhmann, 1990). Hence, what Michel Crozier called the bureaucratic phenomenon, that is, the installation of self-reinforcing mechanisms that lead to an expansion of bureaucracy (Crozier, 2009). The same can be said for the legal system, where laws lead to more laws, lawyers, laws regarding other laws, and court cases (Luhmann et al., 2004). Managing external complexity produces internal complexity, to be managed at a cost, which economists call transaction costs. Stories help seemingly restore simplicity. Stories about evil bureaucracies simply having to shrink do not contribute much to the actual simplification of governance, as they do not acknowledge the underlying reasons for the complexity of governance. Yet many other stories can be utterly helpful, in serving governance and community to understand each other, and themselves, and to guide governance systems in the taking of decisions (Czarniawska & Gagliardi, 2003; Miller, 2012). Stories can assist in the coordination of what has previously been fragmented.

The complexity of governance does lead to coordination issues, and one of the key considerations remaining is incoherence. Single policies can be incoherent, as can the stories inspiring them, the connections between the policy and the story, but also whole policy domains, and, most logically, different policies might start from or lead to contradictory ideas (Peters & Savoie, 1996). Sometimes, incoherence becomes visible in society through implementation of a policy. Incoherence can be innocent, and it can be truly problematic, as when carefully crafted nature conservation policies are jettisoned by ad hoc farming policies, or when human rights legislation is undermined by paranoid border policies.

Technical reasons for incoherence are usually unintentional. There is the process of specialization and compartmentalization already mentioned, and the process of the temporal structuring of a process, where over time, in complex policy processes, different actors, different perspectives, and different motivations and values get involved. Each step in implementation leads to modification and reinterpretation (Flyvbjerg, 2002). In practice, several lines of implementation coexist, increasing the chance of a double incoherence: between subgoals, and between an initial idea and what actually happens. Limited skills, understanding, and resources can represent limits to cohesive policy as well.

It is possible that incoherent policies are accepted because no alternative was available. That is, the policy can be a compromise, or it must coexist with something that is not compatible, a different policy which cannot be abolished. Expedience then dictates incoherence, which can be muffled, or not. Complaints about the issue can be used to split a ruling faction, or to move up on the ladder within leadership. In complex governance systems, it is very well possible that the coexistence of incompatible policies is the result of a lack of knowledge in politics or administration, an ignorance that can be used strategically by or against newcomers. Incoherence can also benefit those in power. Regulatory systems relying on a multitude of specialized roles and on long procedures of implementation, where other policies are encountered, are more prone to this incoherence borne out of ignorance (Desai, 2018). As with anything, ignorance can be feigned, and one person's sincere ignorance can be strategically used by someone else (McGoey, 2012).

While the causes of policy incoherence are more numerous than what we just sketched, the rule of thumb might be clear: more complex governance systems tend to produce less coherent policies (Little, 1997). Now, we would like to discuss the more strategic uses of incoherence, some of which were already intimated. While few actors would benefit from a total collapse of the system, few actors also directly benefit from a perfect functioning, and from the nonexistence of internal contradictions. Thus, some might see advantages in allowing for contradictions, and might strategize toward their persistence (Pusca, 2007). The benefit can reside in a policy that cannot be exposed as incoherent, or it could be associated with the contradiction itself (Howlett & Ramesh, 2014). Strategy can focus on the non-resolution of a contradiction, or on the articulation of a new policy, knowing that issues of coherence might have to be kept silent (Seim & Søreide, 2009). The non-revelation or downplaying of contradiction can produce a rhetoric of open denial, or it can turn into a rhetoric of silence. In parallel, one might want to shift the discussion elsewhere, or highlight the benefits of one of the mismatched policies.

Benefit can accrue from non-implementation, where the incoherence itself might be deployed, or a poorly devised policy brought into incoherent existence. Non-implementation can be advertised or not; lame policies can bring political advantage. One can pretend to support a policy, knowing that implementation is not desirable, and that incoherence is one of the ways to incapacitate it. Politicians, but also advising experts, civil society groups, and actors in administration can avail themselves of this prevalent stratagem. The relation between policy and story, and between front stage story and backstage story, thus comes to the foreground. Insider knowledge is of the essence.

Indeed, in governance, sincerity is not always rewarded, as Machiavelli, and long before him Aristotle knew (Mansfield, 1998). This means that several stories might coexist, that there might be several motivations for supporting a policy or a decision. The distinction can be between a private and a public motivation, involving two stories, but other patterns can easily be imagined. One story can have several motivations; one driving force can engender several stories. One can deploy different rhetorical strategies to convince different stakeholders, and one can shift between stories over time, in adapting to a situation, not knowing that old versions linger on (Urbinati, 2000). One can, moreover, believe in several stories at the same time, oscillate between different explanations for a policy embraced, or, one can overlook the incoherence of a story underlying a policy (Bush, 2012).

If one is sincere and is supporting a policy for one consistent reason that aligns with one story about the good the policy would do, even then incoherence might arise. Translations and interpretations take place in the process of policy articulation and implementation, processes discussed earlier. This means that not every translation of story into policy will work, even if the story itself is coherent, even if there is no resistance in governance that would necessitate modification (Stone, 2012). Translation is in fact decision-making, and decisions never follow unambiguously from their antecedents. Some decisions in the crafting of a policy, will make it harder for others to recognize the story in the policy, and others might create a policy where internal contradictions are perceived, which leaves place for interpretations of insincerity, deal-making, or confusion (Beesley & Hawkins, 2022).

We have discussed the variety of forms and uses of incoherence rather in detail, because it reveals the web of possible connections between policy and narrative, and the variety of strategic opportunities offered by complex governance systems (Crow & Jones, 2018; Eberle, 2019). Incoherence is a product of intention and system reproduction, of knowledge and ignorance, and its uses can be as varied as the intentions and strategic opportunities of actors in governance. A governance system allows for a broad pallet of attitudes toward incoherence, while the motivations to accept or use it, have more sources than usually admitted. Indeed, technical explanations for incoherent governance, or for specific cases, can always be discovered. When ambitions are high, when policy coordination toward greater goals and longer-term strategies is required, existing problems of incoherence might be encountered more directly.

Technical reasons might be real enough, but that does not mean they explain the whole situation. Strategic uses of incoherence are most likely to be found in the darker corners of the governance system, while their potential use can look attractive for those in the know, waiting for the strategic situation to make it profitable. In addition, the rich pallet of uses of incoherence can be linked to the overdetermination of policies. Policies are enacted for a variety of reasons, even if they do not represent a compromise. Those reasons can exist within one actor, and they are not always compatible. Stories are used for a variety of reasons in governance, consciously or not, compatibly or less so. As the relation between policy and story is multilayered itself, open to shifting interpretations, ambivalence toward policy and story results in incoherence and virtually an infinite number of positions toward that incoherence.

In some cases, ambivalence toward the governance system itself, or those in power, can enrich the picture (see already Mitchell, 1959). What can be at stake is power, access to resources, prestige, identity, belonging, and a variety of positive and negative affects (see also Chapter 9). The plural processes of interpretation and translation in governance open up possibilities for others, but also for the initiator to reinterpret the policy, the intention, the goal, and the stories associated with it (cf. Foucault, 2003, 2012). The reproduction of governance itself offers new opportunities all the time for reconnecting policy and story, for reinvesting policy or story with affect, for regarding allies and foes differently. As everything in governance takes place against the horizon of impact on the community, power is always at stake (J. E. Newman & Clarke, 2009).

An investment in a policy or an underlying story tends to be overdetermined, and that overdetermination is not all that transparent. The set of reasons and motivations to embrace a story can easily shift in the policy process, which can be bruising and bothersome, and which can reveal new features of governance and community previously not accounted for. Feelings about a story, its interpretation in policy, tend to become more complicated. Contradictory expectations, unrealistic expectations with actors in governance are likely to stem from overdetermination, unconscious motivations, and from ambivalence.

Incoherence in policy and governance can thus be the result of unacknowledged motivations, of silenced stories, and of competing feelings, yet the same factors can also lead one to overlook incoherence (Glynos, 2010). One might desire to reconcile what cannot be reconciled, to wish differences and fissures out of existence, to overlook limitations in what can be achieved through policy. A desire for harmony, for unity, might inspire all involved to pretend things are coherent, that a grand plan is entirely realistic (S. Newman, 2004; Tomšič & Zevnik, 2015). Weaker arguments might be missed, opponents might be ignored, and counterstrategies wished away.

Both incoherence and coherence can be deployed in actor strategies and collective strategies. Both can be downplayed and exaggerated. Stories of fictional unity, of perfect consensus and shared intention, as we know, can function as productive fictions, possibly bringing coordination into being, and allowing for Voice to emerge (Gunder & Hillier, 2016; see also later this chapter). Ascriptions of unity where it does not exist can also have very different effects, and lead to paranoia with insiders and outsiders, participants and observers (Žižek, 2002).

Ambivalence

It is against this background, of impossible coherence, of strategic use of coherence and incoherence, of overdetermination of motivation and continuous reinterpretation of policy and story, that we need to place the functioning of ambivalence.

Ambivalence, for us, can refer to the oscillation between affective polarities described earlier, of love/hate relations to others, to stories, to ideas about the community and its common goods. It can also refer, more broadly, to the incoherence, sometimes contradictions, that can hide between shifting meanings of and feelings about policies and everything in governance. Ambivalence in the first sense does not have to be at work, in order for ambivalence in the broader sense to be present. Oscillation between polarities stems from a coexistence of those polarities, while ambivalence stemming from over-determination does not need to imply polarities. Incoherence in policy and governance does not have to be the expression of ambivalence, in the first or the second sense, but it can be.

Ambivalence, like incoherence, cannot be banished from governance, just as it cannot be banished from life (Swales & Owens, 2019). What can be achieved in life is limited, but what can be hoped for and aspired to, and what can be desired, much less so. Collective action would be more impactful in the presence of perfect consensus and coordination, but this cannot be expected (Flyvbjerg, 1998). As governance needs to take decisions that will bind and form the collective, the undecidability of those decisions can appear magnified. The implications of what is to be decided come with a burden of responsibility, with self-questioning and anxiety lurking in the shadows. What we encounter in those shadows is ambivalence, not incoherence.

The process of governance itself, the confrontation with opponents and with other stories, the experience of disappointment and attacks, but also with unexpected success and support, and with the reading of different stories and motivations in one's proposals, can all contribute to an amplification of ambivalence (Walker & Shove, 2007). High ideals can be questioned, and new ones can be born (Mannheim, 2013). Confrontation with others, their desires, and their idea of the good community and what governance can mean for it, can bring out new shades of self-reflection that can elucidate new pros and cons, new assumptions and implications of the policy goal or the policy tool (Mansfield, 1998; Thomsen et al., 2011). These insights can resonate with ambivalences already present. Reflection can bring them closer to the surface, or can reshuffle the affective landscape surrounding the policy (Åkerström, 2006).

Ambivalence can be suppressed to various degrees by governance systems and their communities. It can lead to repression, which then engenders symptoms that might be amenable to analysis. Where repression took place, it does not have to hold sway forever, and what was inaccessible can often be unearthed, through modes of learning and reflexivity present in the system, in politics or administration, or through lively public discourse, active civil society and, who knows, maybe even public intellectuals. Where these mechanisms do not work, are absent, or are ensnared in the same repression that created the symptoms, a more intense form of self-analysis in governance might be advisable, a form we present in our final chapter.

New symptoms might appear, new repressions. People revert to old beliefs; new situations trigger regression or new repression. What could not be handled before might be integrated into collective discourse, but new issues might appear that are harder to deal with (Pohl & Swyngedouw, 2023). Governance, by definition an impossible task to merge what cannot be merged and represent what cannot be represented, is bound to be traumatizing, for those involved and for the community, more so than most individual lives. New problems, new disagreements, new tests of cherished narratives and of Master Signifier are never far off, and the reasons for continuous collective and individual repression are always there (Stavrakakis, 2002). Fantasies are produced but are never safe for a long time; identifications are under duress (Glynos, 2010). Pressures to adapt continuously are high as are pressures to ignore, deny, disavow, and repress, especially for those who experience firsthand what is required for governance to function, for it not to slide into the short-term management of practicalities (Medovoi, 2010).

Reducing governance to practical reason, to problem-solving or service provision does not make these complications go away. A radical simplification of the understanding of governance will not reduce ambivalence, the constant movement of desire around ideas of public and individual goods associated with governance (Okulicz-Kozaryn & Valente, 2018; Saari, 2022). Often, a transference of dissatisfaction takes place, from self or others to governance itself, or to those who are unwilling to see what governance really ought to do (Desmet, 2022). In practice, new problems, new hopes and desires do turn up, events that cannot be interpreted and managed within existing frames, new disaffections that are hard to give a place, new and sentiments that cannot be explained, topics which cannot be intellectualized, or affectively quickly neutralized.

New words, as in new concepts, explanations, or a new rhetoric, often do not do the trick, for the intended audience in the community, and for the actors in governance themselves. Neither knowing nor saying guarantees a shift in desire, a reinvestment in governance that might stabilize it. New policy tools or modes of organization are not necessarily helpful either. Collective desire cannot attach forever to one mode of organization, and what might have represented the core values of the community, can appear as drudgery, or worse, can feel like disappointment. This then becomes a symbol of the inadequacy of self, governance, or society itself; it can inspire ambivalence as well as oscillation. One can find positive openings here for radical politics, maybe even a confrontation with the true nature, the full potentiality of politics, à la Alain Badiou, but, too often, disillusion and ambivalence lead to mere destabilization or polarization (Kapoor, 2005; Margulies, 2022).

The act of organization is an intervention in collective desire (Czarniawska, 2008). Ambivalence is bound to remain, to return, as new modes of organization counting on or aiming at collective desire will change that desire, and as hopes for producing collective desire through discourse or organization are never rational. It might work, or not, but what might function as an *objet petit a* for the community might not be clear for those trying to please the community (Fink, 1999). If in the other direction, the community summons governance to create a world it desires, reading that desire, and translating it into something recognized as desirable, leaves plenty of spaces for ambivalence (McGoey, 2012; McGowan, 2019). The self-difference of the community, the gap between community and the governance system it created and is created by, does not allow for stable interpretations and desires.

What retains an affective investment in a policy can be something that is not in the policy itself, or, if it is, it might be something seemingly minor or superficial. It could involve conditions in which the policy was first imagined, or involve the situation the policy was supposed to produce, independent of its reality. Awareness of this conditioning is rare. Supposedly rational decision-makers might be more attached to symbols of rationality than to a quality of reasoning. Desire can attach itself to anything, and, if an *objet petit a* is to be discerned, in the decision-making situation, in the policy, or in the goal, this will not hold the attention forever, whereas its seeming attainment can make it vanish (Dean, 2009).

Small things, small disappointments, and small changes not recognized as disappointment, can shatter an ideal, or break the belief that a policy might help in making it reality. The *objet petit a* moves away, and what was idealized proves normal. Alternatively, when desire did not circle around a policy, yet affect reconnects via new stories, via new encounters, with an object in governance, this can feed a tamed ambivalence, bringing what was stabilized back into oscillation. "Nature" might not look as innocent after farmers' protests. Elsewhere, an unrequited love for an ally is disappointed, and turns into hate, which attaches to anything touched by that ally (Luoma-aho, 2015). When love and desire, hence identification and transference are involved (Heron, 1990), dramatic shifts in affect, goal reversals, and abrupt reinterpretations can be expected. Narcissism, perhaps fed by narratives of self-aggrandizement and mutual congratulation, and narratives initially supporting valuable collective goods and long-term goals, will make things worse and move the needle decidedly from love to hate (Freud, 2014; Frosh, 2016).

What is felt as real and what is worthwhile can alter in the same movement of ambivalent oscillation.

Fear turns into anger, anger attaches to an object, an idea, an ideology that offers itself up in opposition. Disappointed love morphs into hate. Anxiety crystallizes around opponents, real or imagined, presenting themselves as suitable objects. Ambivalence can fuel polarization, and polarization can be driven by fear, for the disruption of an imaginary order, unity, or purity (Gish, 2023; McGowan, 2013, 2017). Once politics becomes identity politics, and polarization is political currency, second-order observation becomes harder. Reflexive and dialectical learning occur less frequently, while suspicion of anything coming from the alienated other, feeds into ambivalence (Stavrakakis, 2008). Trust becomes a scarcer good, and learning what and who could be loved turns into an arduous process. One common response is a partisan closure of the mind, a scorn for any alternative voices, a refusal to engage in meaningful debate. A narcissistic world offers a home for those withdrawing from actual politics, yet even this withdrawal is painful, as the feeling remains that the whole community *ought* to be as imagined (Renström et al., 2023).

Legacies of conflict, trauma, and the Real

An encounter with the Real in governance, is no pleasant matter. The Real encountered is all too often the Real of trauma, of that what could not be symbolized, and haunts the community in its attempts to shape itself, by imaginary and symbolic means (Alpert & Goren, 2016). Hitting the same obstacle, repeating the same moves, clinging to the same unattainable fantasy, obsessing over details that are identified with an order under threat, can all signal repression and trauma. Symptoms might be ways to manage trauma, a precarious balance, yet when these symptoms become severely self-limiting, a democratic process and a community are in need of repair (Rappaport & Simkins, 1991). When we speak of communities, it is worth contemplating whether an unearthing of the repressed, an encounter with the Real in this sense, might not be worth the risk of the de-stabilization it likely entails. Not doing so affects the lives of many and deprives them of the possibility to construct a community as imaginary home, and as a framework in which to pursue their own dreams (see chapter 11)

When what is encountered is the Real as the lack around which desire circles, this can only be traumatic. Fantasies ground our identities, and the governance systems that can reproduce collective identities. Traversing those fantasies, around which so much is structured, unmasking them, is unmasking ourselves as discursive and desiring beings (Vanheule, 2016; Vanheule & Verhaeghe, 2009). Traversal of foundational fantasies is confrontational, disruptive. According to Žižek, it can also be productive, as it opens spaces for real choice, for pure contingency (Bryant, 2016). Such ruptures can thus both be traumatizing and creative, as testified by societies experiencing revolutions, natural disasters, and wars (Bistoen, 2016).

Desire, for Lacan, always assumes a lack at its center, a lack that is constitutive, as without the lack, desire would stop moving, and would cease to be desire (Van Haute, 2002). Communities, their unconscious desires, are similarly built around fantasies, often ideological in nature, which, on close inspection, exert their fascination because their core is empty (Žižek, 2019). When they are directly confronted with this negativity, however, communities can crumble. Governance systems do not need to succumb to this fate, unless they are tightly coupled to the idea of community. Governance represents desire but can also represent arenas and moments where desire can be interrogated, and where the repressed can come to the light. This, of course, entirely depends on Master Signifiers of governance itself, on the functions we allow it to have, on the flexibility and uncertainty we can tolerate as a community, regarding the stability of our self-definition and the value of self-examination.

The collective Superego, which is supposed to guide communities toward their self-declared values, and which can be taken as a symbol for an imagined better future self, is not, still following Lacan, a protection against a destabilizing collective and individual desire or drive (Dean, 1991). Rather, desire is to be seen as a protection against jouissance, transgressive enjoyment that does not serve self-preservation, and is associated with the Real of the drives (McGowan, 2022) (see Chapter 9). The Superego itself, located in both the conscious and the unconscious, is what knows no bounds, and gives itself easily to aggressive drives (Žižek, 2009). Where the Superego pushes for enjoyment of ideological purity, for an aggressive pursuit of perfection, the Real of the drives is at work, and not much good can be expected. Enjoyment of self-destruction can follow enjoyment of a knowingly impossible pursuit of utopian ideals, whether they are of an overtly moral sort, or more conspicuously political. The substance of the injunction and the ideal make a difference, as the promise of heavenly enjoyment makes it easier to accept, or enjoy, the transgression committed in the name of the ideal.

Trauma thus easily creates conflict, and conflict often leads to trauma, while steering for a psychoanalytically inspired traversal of fundamental fantasy comes at great risk. Encounters with the Real, in all its manifestations, come at great risk. Becoming aware of unconscious desires and becoming conscious of the repressed kernels of experience that resist symbolization does not have to come at such cost (see Chapter 11). Similarly, grasping the destabilizing power of aggressively promoted ideals is worth the price of self-analysis.

Histories of conflict and trauma destabilize the community in ways it cannot fully comprehend and manage (Mohatt et al., 2014). Trauma can reassert itself regularly, in symptoms not easily recognized, and easily mistaken for the problem itself. Conflict can polarize politics, and render politics into identity politics, which then limits the possibilities for stabilizing and adapting the community (Van Assche, Gruezmacher et al., 2022; Van Assche et al., 2023). Trauma can become a negative outline, an emptiness with hardening boundaries, and can be transmitted over many generations. Nothing inside is articulated, yet the void exerts a structuring power by pushing away from that which needs to be avoided (Brooks, 2016).

Legacies of conflict and trauma can become problematic for the cohesion of governance, and for the implementation of strategies to move the community in a desirable direction, as they can make it harder to reflect lucidly on anything (Bryant, 2016). Self-observation is rendered more challenging. Reconstructing its own evolution and discerning viable options for future development look more daunting. Blind spots that are very hard to mend, hide issues that need to be addressed, issues which can range from internal relations to relations with the environment (Rust, 2008). Acknowledging that no reconstruction of history provides a vantage point that can show objectively how the past shaped the present of governance, its internal contradictions and imperfections, and its observation of itself and its environment, we nevertheless arrive at a point where careful reconstruction of governance paths is needed, not only in cases of communities traumatized to the point that they are unable to envision any alternative future, but in all situations where a significant change is hoped for. Hence, the next chapter will present a perspective on governance paths and their reconstruction, which can support forms of self-analysis and visioning, for which we can use the label community therapy (see Chapter 12).

Productive fictions

Productive fictions are productive and they tend to be necessary. Governance and community life would cease to exist without them. Fantasy keeps our reality together, and in the world of governance, this is acutely felt, as literally nothing can be taken literally. Stories about good governance, about the past and future of the community must be stories to function, and decisions must be decisions, which means that they have to find support in stories. Nothing would happen if everything were endlessly scrutinized, if politics and administration pondered all day what the community actually is, who represents what, and which assumptions on the good life and the real economy are built into their daily routines (Alvesson & Spicer, 2016). Moreover, without the glue of fantasy, conflicts would erupt, as the binding power of Master Signifiers would erode.

Productive fictions are useful in glossing over incoherence. They can quiet down ambivalence or move it to the background. They can assist in the non-observation of incoherence, in the suppression of ambivalence, by simply looking away, or by directing attention to what is supposed to be true (Žižek, 2002). Abandoning productive fictions can cause anxiety, for individuals and for groups. Within governance systems, which never work as they believe themselves to work, this must be avoided, hence the institutionalization, repetition, and ritualization of productive fictions (Bistoen, 2016). This institutionalization can lead to functional stupidity, under the flag of Master Signifiers such as efficiency, democracy, participation, transparency, and others. Such signifiers cannot be missed. Where they become less productive, is where learning infrastructures are affected, which are necessary for adaptation to always changing circumstances (Van Assche, Duineveld, et al., 2022). Hence, the succession of Master Signifiers in governance, and the never-ending waves of administrative reform which are expected to "clean the house" that was apparently always dirty (Czarniawska-Joerges, 1989; Farmer, 1995; J. E. Newman & Clarke, 2009).

Productive fictions thus assist in avoiding and evading that which cannot be acknowledged. That which cannot be recognized can be incoherence and ambivalence, but it can also be stories, feelings, and histories that cannot come to the surface because they would conflict with ruling ideology, morality, or power relations. Affects can resurface in unexpected places when productive fictions do not succeed fully in keeping the repressed at bay (Massumi, 2015). Stories can remain latent, yet feelings associated with them might attach to new topics, linked through similarities recognized in the system, sometimes opening up those topics for discussion from new angles, and elsewhere closing them off.

Productive fictions are not eternal, might be challenged in governance, and are themselves not fully compatible with each other. Each ideology has internal consistencies, and, following Žižek, an empty core, which if revealed, dissolves the elements and removes its persuasive power (Žižek, 1992, 2019). Changes in the latent infrastructure of productive fictions can stem from new leadership, from cultural change, and from policy failures. An internal diversity of perspectives, a tolerance for uncertainty, resistance against pressures to simplify and streamline, and the cultivation of self-critique build resistance against the immunization of governance against learning (Van Assche et al., 2024). Learning processes in governance help in moving alternative explanations, identities, and policy options into the background, in shifting the pattern of productive fictions, without leading to repression.

Productive fictions do more of the heavy lifting in governance when ambitions are higher, when communities have to convince themselves that hard work and some self-denial, toward a better future are worth it. At the same time, they do not stand in the way of pragmatism, in navigating the governance system, in the practicality of everyday implementation and service provision. They do not hamper the strategizing of actors making their way up the ladder, and they do not necessarily lead society into a world of fantasy. In fact, stories about good governance, about the power of governance, and about the possibility of perfect representation and participation can be regarded as productive fictions. By building them into other stories about governance, and by assuming them in strategies that pursue public goods in an ambitious manner, much more can be achieved than by systems where self-limiting assumptions are the cornerstone of operational identity (see Chapters 10 and 11).

De-paradoxification

Now, we would like to pause for a moment and introduce Niklas Luhmann (1927–1998). Luhmann was a German systems theorist, a brilliant interpreter of the relations between politics, law, and administration, between governance and society. He was not a psychoanalyst and did not discuss the human psyche, which he believed to exist in the environment of social systems (Luhmann, 1995). Yet, when we investigate how psychoanalysis can shed a light on governance, we encounter a series of utterly useful insights, which sometimes parallel psychoanalytic reasoning and sometimes complement it.

For Luhmann, organizations operate on the basis of impossibilities, of paradox (Luhmann, 2018). In fact, this also applies to the more encompassing social systems of law, politics, and administration, and it applies to society as a whole (Luhmann, 1995). These systems are continuously faced with situations where the conditions of their possibility (or truth) are also the possibilities of their impossibility (or untruth). It is thus not really a matter of being true and not true at the same time, but of *observing* something where the distinction made enables contradictory interpretations – as observation is making distinctions (Brooks, 2016). The environment, for a system, according to Luhmann, is always an internal construct, as what is out there, is only there for us as observed by social systems, which are not produced by what is out there but by people. Systems build models of relevant environments according to their own operations (Luhmann, 1989). Paradoxes are not necessarily a problem for systems but rather for observers. However, when observers in systems directly confront a paradox, this can paralyze them as participants in the system. Thus, law cannot answer the question directly as to why something is a law, except by shifting the answer to procedures (away from substance) or by producing a tautology (the law is the law). Both sides of a distinction remain in existence, enabling a reversal of the distinction, a return of the repressed, so to speak.

For Freud, the unconscious knows no negatives, and both an idea and its opposite can be treated in the same way, forming associations, becoming displaced, condensed, and generating symptoms (Freud, 2020). One can interpret this as a weakening, not of the distinction (as one can still recognize two things produced by that particular distinction) but of the highlighting of one side that takes place in conscious life, and hence in the communications that build up social systems. For Freud, the weakening of the controls of consciousness (as in sleep) allows not only for the surfacing of what could not be acknowledged for oneself, but also the facing of paradox, setting each side of the distinction to work at the same time.

What governance systems need to do, as they cannot fall asleep too often, is not face the paradox but hide it. For organizations participating in governance, and for the system as a whole, this means that they cannot admit there is ultimately no argument, no solid ground for what they are doing, for the general direction taken, or for the connection between that direction and everyday decision-making (Kornberger, 2022; Seidl, 2007). Productive fictions are required, and now we can say that one of their functions is de-paradoxification. They are part of strategies to hide the fundamental impossibility of governance, in its self-declared form, in its own and the public's imaginary. There is no real reason for it to do what it does, and there is no way it can literally do what it literally promised. Certainly, there are always answers to the why question, and explanations for the difference between promise and outcome, but those will ultimately circle back to the productive fictions and other rhetorical devices that are part of the repertoire of de-paradoxification (cf. Beunen & de Vries, 2011; Pressman & Wildavsky, 1984). Meanwhile, the nature of politics and communities makes it nearly impossible to honestly reflect how governance works, as promises need to be made, simplified identities honored, impossible expectations somehow managed, and uncertainties underplayed.

De-paradoxification in the daily life of an organization comprises many things, including the ascription of necessity to contingent decisions, by constructing urgency and by devising stories about events in the environment that are supposed to force its hand (Beunen & Van Assche, 2021). Stories about self and environment can reinforce each other to determine an action. Let's say we are slow and people tell us we are slow, so we must be fast, since fast is efficient and efficient is good. Temporalities are constructed that can function as a reason for decision and action: we need to do something now, as action is important, so people can see something and believe we are useful. Or: we need to approve more permits, as targets are not being met and the minister might be changed. What counts as an argument refers to stories that refer to other stories which are ultimately made up by the organization itself, or by others in the governance systems, allowing for a combination of organizational and system-wide de-paradoxification (Stetter, 2005). Our earlier observations on incoherence appear in a slightly different light here, as some cases of the coexistence of incompatibilities as well as the playing out of ambivalence, can be connected now to strategies of de-paradoxification, of carefully, even if unconsciously, avoiding the paradox.

Avoiding the paradox is avoiding confrontation with the lack of essence and the lack of foundation in the organization, the community, and the individual (Fuchs, 2001). We build up our societies out of individuals and organizations, and build up individuals out of stories stemming from societies and, indeed, organizations. Nobody has an objective reason, and an external grounding, for their own identity. Environments do not dictate the functioning of systems nor the identity of individuals. Referring to free choice, as the other side of the distinction, is not sufficient. What takes place is contingent co-evolution, where options always exist, where different development paths are always possible, guided by a web of stories telling us who we are and what would be a reasonable or meaningful decision.

For Lacan, this likely comes as no surprise. Freud, too, might not see much new, as for him too, the unfolding of individuality is a process of contingent encounters leading to structures and processes that then interpret future contingent encounters. This happens in ways that suggest identity to system and observer, to person and environment. Desire might, for Lacan, stem from the unknowability of the world, from the necessary distance between self and the world that is mirrored in the self, in the split subject, yet this does not make desire autarkic (Eyers, 2012; Kapoor, 2005; Margulies, 2022). It borrows signs and symbols, stories and their affects, from others, from the cultural and political world, and it derives its power both from those signs, and from their imperfection, from their inability to capture what one feels and what one wants. Both the what and the why of desire borrow from the world of signifiers available to us and encountered by us, yet that world never succeeds for longer than a moment, to tell us what to desire.

Concluding

Governance is imperfect, in many ways. Real existing systems of governance, instantiations of ideas regarding good governance, and adaptations to practical realities of limited resources, ideas, coordination tools, and support make for a multiplicity of modes of imperfection. In this chapter, we examined imperfection through the lens of incoherence and ambivalence, which were found to have positive and negative functions and effects, prevalent in their intentional and unintentional forms.

We revisited the idea of productive fictions in governance, highlighting not just their productivity but also their necessity. Without fictions, without fantasy, communities cannot exist and governance systems cannot help them come into existence and transform themselves. We formally introduced Niklas Luhmann, already waiting behind the scenes, whose central place for paradox in the organization of social life meant that people, organizations, and societies had to evolve a toolbox to ignore, deny, evade, displace, and sublimate the paradox, a toolbox for de-paradoxification. Not every productive fiction supports de-paradoxification, and not every example of de-paradoxification relies on productive fictions, but their combined relevance for governance is no coincidence. They help explain the roles of incoherence and ambivalence in governance, and they can claim a central place easily in a system on which ideals are projected from all directions, where blame is placed for virtually anything, and where the tragedy of the self-splitting between governance and community leaves marks of disappointed expectation every day.

If we emphasize contingency and co-evolution, it makes sense to reconstruct the history of governance systems, as its unique modes of de-paradoxification, of recognizing and building identities, of self-regulation through meta-narratives can only be understood by tracing them in that history. What is forgotten can be just as important as what is remembered, and how things are remembered just as meaningful as what is remembered (Fink, 1999, 2017). A very practical implication is that system-specific modes of defining and ignoring problems can be more telling than mapping the repertoire of problem-solving modes (Ostrom, 2008). Finding inspiration again in psychoanalysis, it makes sense to see reconstruction of what we call governance paths as belonging to the realm of analysis, and to see the driving force of such analysis as the governance system itself. Analysis has to be first of all self-analysis. We develop this idea in Chapter 11, but first things first. Let us now investigate the nature of governance paths, the paths shaping that which later must analyze itself.

References

Åkerström, M. (2006). Doing ambivalence: Embracing policy innovation—At arm's length. *Social Problems, 53*(1), 57–74. https://doi.org/10.1525/sp.2006.53.1.57

Alpert, J. L., & Goren, E. R. (2016). *Psychoanalysis, Trauma, and Community: History and Contemporary Reappraisals*. Taylor & Francis.

Alvesson, M., & Spicer, A. (2016). *The Stupidity Paradox: The Power and Pitfalls of Functional Stupidity at Work*. Profile Books.

Beesley, C., & Hawkins, D. (2022). Corruption, institutional trust and political engagement in Peru. *World Development, 151*, 105743. https://doi.org/10.1016/j.worlddev.2021.105743

Beunen, R., & de Vries, J. R. (2011). The governance of Natura 2000 sites: The importance of initial choices in the organisation of planning processes. *Journal of Environmental Planning and Management, 54*(8), 1041–1059. https://doi.org/10.1080/09640568.2010.549034

Beunen, R., & Van Assche, K. (2021). Steering in governance: Evolutionary perspectives. *Politics and Governance, 9*(2), 365–368. https://doi.org/10.17645/pag.v9i2.4489

Bistoen, G. (2016). The Lacanian Concept of the Real in Relation to Politics and Collective Trauma. In G. Bistoen (Ed.), *Trauma, Ethics and the Political beyond PTSD: The Dislocations of the Real* (pp. 104–130). Palgrave Macmillan UK. https://doi.org/10.1057/9781137500854_6

Brans, M., & Rossbach, S. (1997). The autopoiesis of administrative systems: Niklas Luhmann on public administration and public policy. *Public Administration, 75*(3), 417–439.

Brooks, R. M. (2016). The intergenerational transmission of the catastrophic effects of real world history expressed through the analytic subject. In J. Mills & R. C. Naso (Eds.), *Ethics of Evil* (pp 137–177). Routledge.

Bryant, L. R. (2016). Symptomal knots and evental ruptures: Žižek, Badiou, and discerning the indiscernible. *International Journal of Žižek Studies, 1*(2).

Bush, R. (2012). Rhetoric, psychoanalysis, and the imaginary. *Cultural Studies, 26*(2–3), 282–298. https://doi.org/10.1080/09502386.2011.644115

Crow, D., & Jones, M. (2018). Narratives as tools for influencing policy change. *Policy & Politics, 46*(2), 217–234. https://doi.org/10.1332/030557318X15230061022899

Crozier, M. (2009). *The Bureaucratic Phenomenon*. Transaction Publishers.

Czarniawska, B. (2008). Organizing: How to study it and how to write about it. *Qualitative Research in Organizations and Management: An International Journal, 3*(1), 4–20. https://doi.org/10.1108/17465640810870364

Czarniawska, B., & Gagliardi, P. (2003). *Narratives We Organize by* (Vol. 11). John Benjamins Publishing.

Czarniawska-Joerges, B. (1989). The wonderland of public administration reforms. *Organization Studies, 10*(4), 531–548. https://doi.org/10.1177/017084068901000404

Dean, J. (1991). Still Dancing: Drive as a Category of Political Economy. In H. Feldner & F. Vighi (Eds.), *States of Crisis and Post-Capitalist Scenarios* (pp 59–71). Routledge.

Dean, J. (2009). Politics without politics. *Parallax, 15*(3), 20–36. https://doi.org/10.1080/13534640902982579

Desai, D. (2018). Ignorance/Power: Rule of Law Reform and the Administrative Law of Global Governance. In M. Hirsch & A. Lang (Eds.), *Research Handbook on the*

Sociology of International Law (pp. 151–188). Edward Elgar Publishing. www.elga ronline.com/edcollchap/edcoll/9781783474486/9781783474486.00015.xml

Desmet, M. (2022). *The Psychology of Totalitarianism*. Chelsea Green Publishing.

Duit, A., & Galaz, V. (2008). Governance and complexity—Emerging issues for governance theory. *Governance, 21*(3), 311–335.

Eberle, J. (2019). Narrative, desire, ontological security, transgression: Fantasy as a factor in international politics. *Journal of International Relations and Development, 22*(1), 243–268. https://doi.org/10.1057/s41268-017-0104-2

Eyers, T. (2012). *Lacan and the Concept of the "Real."* Palgrave Macmillan.

Farmer, D. J. (1995). *The Language of Public Administration: Bureaucracy, Modernity and Postmodernity*. University of Alabama Press.

Fieschi, C., & Heywood, P. (2004). Trust, Cynicism and Populist Anti-politics. *Journal of Political Ideologies, 9*(3), 289–309. https://doi.org/10.1080/135693104200 0263537

Fink, B. (1999). *A Clinical Introduction to Lacanian Psychoanalysis: Theory and Technique*. Harvard University Press.

Fink, B. (2017). *A Clinical Introduction to Freud: Techniques for Everyday Practice*. W. W. Norton & Company.

Flyvbjerg, B. (1998). *Rationality and Power: Democracy in Practice*. University of Chicago press.

Flyvbjerg, B. (2002). Bringing power to planning research. One researcher's praxis story. *Journal of Planning Education and Research, 21*, 353–366.

Foucault, M. (2003). *Society must be Defended: Lectures at the College de France, 1975–76*. Allen Lane The Penguin Press.

Foucault, M. (2012). *Discipline and Punish: The Birth of the Prison*. Knopf Doubleday Publishing Group.

Freud, S. (2014). *The Neuro-Psychoses of Defence*. Read Books Ltd.

Freud, S. (2020). *The Interpretation of Dreams: The Psychology Classic*. John Wiley & Sons.

Frosh, S. (2016). Relationality in a time of surveillance: Narcissism, melancholia, paranoia. *Subjectivity, 9*(1), 1–16. https://doi.org/10.1057/sub.2015.19

Fuchs, S. (2001). *Against Essentialism: A Theory of Culture and Society*. Harvard University Press.

Gish, E. (2023). When purity cannot save us: On matter out of place and democratic hope. *Theology & Sexuality, 29*(2–3), 219–235. https://doi.org/10.1080/13558 358.2024.2332977

Glynos, J. (2010). *Lacan at Work?* (C. Cederstrom & C. Hoedemaeker, Eds.; pp. 13–58). MayFly Books. https://repository.essex.ac.uk/4085/

Green, D., & Shapiro, I. (1994). *Pathologies of Rational Choice Theory: A Critique of Applications in Political Science*. Yale University Press.

Gunder, M., & Hillier, J. (2016). *Planning in ten words or less: A Lacanian entanglement with spatial planning*. Routledge.

Heron, J. (1990). The politics of transference. *Self & Society*. www.tandfonline.com/doi/abs/10.1080/03060497.1990.11085042

Howlett, M. (2000). Managing the "hollow state": Procedural policy instruments and modern governance. *Canadian Public Administration/Administration Publique Du Canada, 43*(4), 412–431. https://doi.org/10.1111/j.1754-7121.2000.tb01152.x

Howlett, M., & Ramesh, M. (2014). The two orders of governance failure: Design mismatches and policy capacity issues in modern governance. *Policy and Society, 33*(4), 317–327. https://doi.org/10.1016/j.polsoc.2014.10.002

Kapoor, I. (2005). Participatory development, complicity and desire. *Third World Quarterly, 26*(8), 1203–1220. https://doi.org/10.1080/01436590500336849

Keating, M., McAllister, I., Page, E. C., & Peters, B. G. (2023). *The Problem of Governing: Essays for Richard Rose*. Springer Nature.

Kornberger, M. (2022). *Strategies for Distributed and Collective Action Connecting the Dots*. Oxford University Press USA - OSO. http://public.eblib.com/choice/PublicFul lRecord.aspx?p=6836917

Little, J. H. (1997). Maturana, Luhmann, and self-referential government: Is democratic administration even possible? *Administrative Theory & Praxis, 19*(3), 342–354.

Luhmann, N. (1989). *Ecological Communication*. University of Chicago Press.

Luhmann, N. (1990). *Political Theory in the Welfare State*. De Gruyter.

Luhmann, N. (1995). *Social Systems* (Vol. 1). Stanford University Press Stanford.

Luhmann, N. (2018). *Organization and Decision*. Cambridge University Press.

Luhmann, N., Ziegert, K. A., & Kastner, F. (2004). *Law as a Social System*. Oxford University Press.

Luoma-aho, V. (2015). Understanding stakeholder engagement: Faith-holders, Hateholders & Fakeholders. *RJ-IPR: Research Journal of the Institute for Public Relations, 2*(1), 62–89. https://jyx.jyu.fi/handle/123456789/45268

Mannheim, K. (2013). *Ideology and Utopia*. Routledge. https://doi.org/10.4324/9781315002828

Mansfield, H. C. (1998). *Machiavelli's Virtue*. University of Chicago Press.

Margulies, J. (2022). A political ecology of desire: Between extinction, anxiety, and flourishing. *Environmental Humanities, 14*(2), 241–264. https://doi.org/10.1215/22011919-9712357

Massumi, B. (2015). *Politics of Affect*. John Wiley & Sons.

McGoey, L. (2012). The logic of strategic ignorance. *The British Journal of Sociology, 63*(3), 533–576. https://doi.org/10.1111/j.1468-4446.2012.01424.x

McGowan, T. (2013). *Enjoying What We Don't Have: The Political Project of Psychoanalysis*. U of Nebraska Press.

McGowan, T. (2017). *Capitalism and Desire: The Psychic Cost of Free Markets*. Columbia University Press. https://doi.org/10.7312/mcgo17872

McGowan, T. (2019). Law and Superego. In Y. Stavrakakis (Ed.), *Routledge Handbook of Psychoanalytic Political Theory* (pp. 139–150). Routledge.

McGowan, T. (2022). Sublimating the commodity. *Distinktion: Journal of Social Theory, 23*(2–3), 359–377. https://doi.org/10.1080/1600910X.2022.2054447

Medovoi, L. (2010). A contribution to the critique of political ecology: Sustainability as disavowal. *New Formations, 69*(69), 129–143. https://doi.org/10.3898/NEWF.69.07.2010

Miller, H. T. (2002). *Postmodern Public Policy*. SUNY Press.

Miller, H. T. (2012). *Governing Narratives: Symbolic Politics and Policy Change*. University of Alabama Press.

Mitchell, W. C. (1959). The ambivalent social status of the American Politician. *Western Political Quarterly, 12*(3), 683–698. https://doi.org/10.1177/106591295901200303

Mohatt, N. V., Thompson, A. B., Thai, N. D., & Tebes, J. K. (2014). Historical trauma as public narrative: A conceptual review of how history impacts present-day health. *Social Science & Medicine*, *106*, 128–136. https://doi.org/10.1016/j.socsci med.2014.01.043

Newman, J. E., & Clarke, J. H. (2009). *Publics, Politics and Power: Remaking the Public in Public Services*. Sage.

Newman, S. (2004). Interrogating the Master: Lacan and radical politics. *Psychoanalysis, Culture & Society*, *9*(3), 298–314. https://doi.org/10.1057/palgrave.pcs.2100021

Okulicz-Kozaryn, A., & Valente, R. R. (2018). City life: Glorification, desire, and the unconscious size fetish. In I. Kapoor (Ed.), *Psychoanalysis and the Global* (pp 209–232). University of Nebraska Press.

Ostrom, V. (2008). *The Intellectual Crisis in American Public Administration*. University of Alabama Press.

Peters, B. G., & Savoie, D. J. (1996). Managing incoherence: The coordination and empowerment conundrum. *Public Administration Review*, *56*(3), 281–290. https://doi.org/10.2307/976452

Pohl, L., & Swyngedouw, E. (2023). Enjoying climate change: *Jouissance* as a political factor. *Political Geography*, *101*, 102820. https://doi.org/10.1016/j.pol geo.2022.102820

Pressman, J. L., & Wildavsky, A. (1984). *Implementation: How Great Expectations in Washington Are Dashed in Oakland; Or, Why It's Amazing that Federal Programs Work at All, This Being a Saga of the Economic Development Administration as Told by Two Sympathetic Observers Who Seek to Build Morals*. Univ of California Press.

Pusca, A. (2007). Shock, therapy, and postcommunist transitions. *Alternatives*, *32*(3), 341–360. https://doi.org/10.1177/030437540703200304

Rappaport, J., & Simkins, R. (1991). Healing and empowering through community narrative. *Prevention in Human Services*, *10*(1), 29–50. https://doi.org/10.1300/J293v10n01_03

Renström, E. A., Bäck, H., & Carroll, R. (2023). Threats, emotions, and affective polarization. *Political Psychology*, *44*(6), 1337–1366. https://doi.org/10.1111/pops.12899

Rust, M. J. (2008). Climate on the couch: unconscious processes in relation to our environmental crisis. *Psychotherapy and Politics International*, *6*(3), 157–170.

Saari, A. (2022). Topologies of desire: Fantasies and their symptoms in educational policy futures. *European Educational Research Journal*, *21*(6), 883–899. https://doi.org/10.1177/1474904120988389

Seidl, D. (2007). General strategy concepts and the ecology of strategy discourses: A systemic-discursive perspective. *Organization Studies*, *28*(2), 197–218.

Seidl, D., & Becker, K. H. (2006). Organizations as distinction generating and processing systems: Niklas Luhmann's contribution to organization studies. *Organization*, *13*(1), 9–35. https://doi.org/10.1177/1350508406059635

Seim, L. T., & Søreide, T. (2009). Bureaucratic complexity and impacts of corruption in utilities. *Utilities Policy*, *17*(2), 176–184. https://doi.org/10.1016/j.jup.2008.07.007

Stavrakakis, Y. (2002). *Lacan and the Political*. Routledge.

Stavrakakis, Y. (2008). Peripheral vision: Subjectivity and the organized other: Between symbolic authority and fantasmatic enjoyment. *Organization Studies*, *29*(7), 1037–1059. https://doi.org/10.1177/0170840608094848

Stetter, S. (2005). The politics of De-Paradoxification in Euro-Mediterranean relations: Semantics and structures of 'Cultural dialogue.' *Mediterranean Politics*, *10*(3), 331–348. https://doi.org/10.1080/13629390500289367

Stone, D. (2012). Transfer and translation of policy. *Policy Studies*, *33*(6), 483–499. https://doi.org/10.1080/01442872.2012.695933

Swales, S., & Owens, C. (2019). *Psychoanalysing Ambivalence with Freud and Lacan: On and Off the Couch*. Routledge. https://doi.org/10.4324/9780429448652

Thomsen, D. K., Tønnesvang, J., Schnieber, A., & Olesen, M. H. (2011). Do people ruminate because they haven't digested their goals? The relations of rumination and reflection to goal internalization and ambivalence. *Motivation and Emotion, 35*(2), 105–117. https://doi.org/10.1007/s11031-011-9209-x

Tomšič, S., & Zevnik, A. (2015). *Jacques Lacan: Between Psychoanalysis and Politics*. Routledge.

Urbinati, N. (2000). Representation as advocacy: A study of democratic deliberation. *Political Theory, 28*(6), 758–786. https://doi.org/10.1177/0090591700028006003

Van Assche, K., Beunen, R., & Gruezmacher, M. (2024). *Strategy for Sustainability Transitions: Governance, Community and Environment*. Edward Elgar Publishing.

Van Assche, K., Duineveld, M., Beunen, R., & Teampau, P. (2011). Delineating locals: Transformations of knowledge/power and the governance of the Danube Delta. *Journal of Environmental Policy & Planning, 13*(1), 1–21. https://doi.org/10.1080/1523908X.2011.559087

Van Assche, K., Duineveld, M., Beunen, R., Valentinov, V., & Gruezmacher, M. (2022). Material dependencies: Hidden underpinnings of sustainability transitions. *Journal of Environmental Policy & Planning, 24*(3), 281–296. https://doi.org/10.1080/15239 08X.2022.2049715

Van Assche, K., Gruezmacher, M., & Beunen, R. (2022). Why governance is never perfect: Co-evolution in environmental policy and governance. *Sustainability, 14*(15), Article 15. https://doi.org/10.3390/su14159441

Van Assche, K., Gruezmacher, M., Marais, L., & Perez-Sindin, X. (2023). *Resource Communities: Past Legacies and Future Pathways*. Taylor & Francis. https://books.google.ca/books?id=4EHUEAAAQBAJ

Van Haute, P. (2002). *Against Adaptation: Lacan's "Subversion" of the Subject*. Other Press.

Vanheule, S. (2016). Capitalist discourse, subjectivity and Lacanian psychoanalysis. *Frontiers in Psychology, 7*. https://doi.org/10.3389/fpsyg.2016.01948

Vanheule, S., & Verhaeghe, P. (2009). Identity through a psychoanalytic looking glass. *Theory & Psychology, 19*(3), 391–411. https://doi.org/10.1177/0959354309104160

Walker, G., & Shove, E. (2007). Ambivalence, sustainability and the governance of socio-technical transitions. *Journal of Environmental Policy & Planning, 9*(3–4), 213–225. https://doi.org/10.1080/15239080701622840

Žižek, S. (1992). *Looking Awry: An Introduction to Jacques Lacan through Popular Culture*. MIT Press.

Žižek, S. (2002). *Did Somebody Say Totalitarianism?: Five Interventions in the (mis) use of a Notion*. Verso.

Žižek, S. (2009). *In Defense of Lost Causes*. Verso. http://public.ebookcentral.proquest.com/choice/publicfullrecord.aspx?p=5176960

Žižek, S. (2019). *The Sublime Object of Ideology*. Verso Books.

8 Governance paths enabling and limiting

Governance paths and dependencies

Psychoanalysis offers insight into both the limiting and the enabling effect of the structures that evolve over time in governance and community, ranging from dependence on narratives to dependence on others, on a preferred past and future, on selective blindness and productive fictions, and on elementary scenarios framing the understanding of new situations. Psychoanalysis further illuminates how limitation only becomes a problem when it impedes everyday functioning, or deeply held values and aspirations, when it leads to breakdowns in governance, failed ambitions, and persistent problems of non-implementation and disillusion, and when it sets communities on a course of fragmentation and persistent conflict (cf. Van Haute, 2002). One can speak of a problem when internal divisions and contradictions become polarizing and fragmenting, when no more interpretations are available that keep the community and polity together. There is a problem when the direction taken leads to oppression, or when it leads to shrinkage in the perspective on the future and its possibilities for further self-transformation (Kapoor, 2015).

If communities, for reasons not wished by them, but resulting from their own partly opaque history, are not able to understand and imagine that various alternative courses of development (and adaptation) remain possible, and are not able realistically to assess the viability and desirability of alternative paths, then that is a problem for whatever form of democracy or political ideology (Ziai, 2004). Making the invisible visible, making the unconscious conscious, remains a valid adage, for individuals and for communities; we develop this point toward the end of the book.

But how should we understand those histories of governance in a way that is more revealing regarding opacities and forms of self-limitation at work? Here, evolutionary governance theory (Beunen et al., 2014) offers answers. For evolutionary governance, a theory of governance that is not normative but analytic, and that incorporates insights from systems theory, post-structuralism, and institutional economics, governance systems cannot be transparent to themselves (Valentinov et al., 2019). They are the product of contingent co-evolutions between actors and institutions, and power and

DOI: 10.4324/9781032696676-8

knowledge, leading to configurations that are cohesive, meaning that they produce decisions and institutions which can guide behavior and decision-making, and meaning that they consist of elements that have to respond to each other, as a result of their co-evolution, and the possibilities and requirements stemming from their goal, that is, the making and implementation of collectively binding decisions (Van Assche et al., 2013).

Actors, for evolutionary governance theory (EGT), can be individuals, organizations, and groups, yet the participation of groups in governance will most likely require a process of formalization and organization, enabling the selection of representatives mandated with a program. Encounters with complexity and self-imposed goals thus lead to representation, to the formation of semantics, programs, and procedures, to the development of internal complexity (Luhmann, 1995, 2018; Luhmann et al., 2004). Organizations have their own identity, guiding their decision-making, interactions with the environment, goals, rules of membership, and the search for resources (Seidl, 2016). Actors have to follow rules codified by institutions, and actors assist in the production of new institutions, which can guide decision-making in governance and in the community.

Those institutions can be simple rules, or more complex institutions, such as policies, laws, and plans. In complex societies, which articulate goals for themselves and that guide a wide diversity of activities, these types of institutions can coordinate others, lean on others, and delineate or constrain the use of others (Van Assche et al., 2020). Plans can be produced because of enabling legislation; their implementation relies on other plans, on policies coordinated, with each policy enabled by different laws. Actors and institutions co-evolve, as actors are able to participate and strategize in governance, thus shaping themselves as actors, through reliance on, production of, and strategic use of institutions, while institutions take shape in a process of use by actors, and by the collective, to solve problems and pursue collective goals. Hence, we can speak of actor/institution configurations, as one side cannot be understood without the other, and as changes on one side will trigger changes on the other side. If institutions are not used, they will lose their teeth, their power to coordinate decision-making and action. That is, actors in governance and people in the community will not be used to them anymore, will not trust them, and alternative tools of coordination will take their place (to solve similar problems)

Similarly, power and knowledge shape each other over time, and following Foucault, we speak of power/knowledge configurations (see also next chapter). Power always entails the creation or deployment of knowledge, as power by mere coercion or simple rule-following rarely suffices (Hillier, 2002; Newman, 2005). Even where rules are blindly followed, some belief in the underlying principles and truths is usually implied. Policies are produced through political and administrative systems incorporating a variety of expert knowledges, including expertise of the system itself, and various stories about problems, public goods, good governance, and amazing looming opportunities. Governance attracts many stories about the community, many versions of reality, because it offers the opportunity, or the promise, to reinforce those realities, to bring desired futures closer, to finally solve those problems apparently overlooked by everyone else. Once a version of reality, through narratives or expert knowledges is codified in institutions, they have a greater chance to become reality for more people in more ways, while chances that opposition can crystallize also increase (Elden, 2017; Pottage, 1998).

The unique way a governance system evolves, the particular history of interactions between subsystems, leads us to speak of *governance paths*. Each system, each polity, and each community organizing itself as a polity, coming into being through governance, is marked by its own path (Beunen et al., 2022; Van Assche et al., 2024). Because certain tools of coordination are more functional and accepted than others, because only certain realities are recognized and certain arguments carry weight, and as coordination is stabilized in patterns of acceptable, recognizable, and functional actors and institutions, the modes to change these patterns, to shift the path, are limited (Braithwaite, 2020). Path creation always remains possible, yet each path comes with its own, formal and informal rules regarding what can be changed, that is, its own ideas and feelings regarding what and when rules should be changed (Simmie, 2014).

We therefore speak of dependencies as rigidities in the governance path, and distinguish path dependencies, interdependencies, goal dependencies, and material dependencies that are at work in every system (Sehring, 2009; Van Assche et al., 2017). Path dependencies are legacies from the past, visible in the structure and functioning of governance now, and we can recognize organizational, institutional, and cognitive path dependencies, where institutional dependencies refer to the world of actor/institution configurations, and cognitive path dependencies to the world of discourse, the world of power and knowledge. Interdependencies are rigidities stemming from preferential relations and tight coupling between actors, between institutions, and between actors and institutions (Duineveld et al., 2013, 2017). Actors can keep each other in place; they deploy a select set of institutions to maintain their positions, while certain institutions are harder to change than others, because of their entanglement with, or solid foundation in, a wide array of other institutions.

Material dependencies are effects of the material world, the physical world, on the functioning of governance. They can manifest themselves in each aspect of governance and can result from previous decisions made in the system, as in the case of large infrastructure works, or they can be caused by features of or changes in the natural environment (Van Assche, Duineveld et al., 2022). Material dependencies can be observable for the system or not so much so (Boin et al., 2020; Duineveld et al., 2017). Goal dependencies, finally, are the effects of visions of the future on the functioning of governance itself. This can be directly, as when a system must realign in order to make a strategy work, or indirectly, as when effects of a previous plan or imagined future in the community come back as positive or negative feedback and affect governance – as when a policy is felt to fail and a backlash follows, or a counter-policy is formulated.

Dependencies are thus limits to the flexibility of governance, in its self-transformation and adaptation, and they represent limits of governance. They shape what is understood as governable, as worthy of and amenable to collective decision-making (Kooiman, 2020). Governance paths shape who is included, what counts as an actor, which tools of coordination are going to be deployed, which versions of reality, embedded values, and aspirations are involved. In governance paths, collective affects are invoked in unique ways.

Self-limitation and limitation

Limits can be changed or accepted. They can be observed or remain unobserved. They can be understood as a weakness or a strength. Sometimes, a boundary in governance is directly imposed by the material or discursive environment, as when things simply do not work anymore, or are simply not acceptable anymore, yet more often, the self-limitation in governance and community that emerges over time comes out of an interplay between all elements in governance. What is important to understand is that the structures and mechanisms that are limiting, are also enabling, as any structure is both (Foucault, 2002b, 2002a). This applies to the worlds of thinking and organizing alike (Czarniawska, 2014). Identity and memory hold power over the interplay of all elements in governance, as that which can be remembered, has a greater chance to play a role in decision-making, and as identity, structuring and structured by memory, guides decision-making, directly, through invocation, or indirectly, through the application of routines associated with it (Benwell, 2006; Brockmeier, 2002; Gabriel, 1999; Luhmann, 2018).

Governance is not merely a reflection of the values of the community, or one part among others, but the subsystem of society that is endowed with legitimacy and power to create or maintain the community in a state imagined as real and desirable by the community. The subsystem develops its own identity and forms of closure, a marked influence on and mediation of the images and desires in the community (Clarke, 2005; Hoggett, 2015). The logic of governance itself, its power dynamics, patterns of inclusion and exclusion, tools of coordination, and forms of knowledge available reproduce limits to thinking and organizing, to understanding and imagining, which then become entangled with the limits emanating from public discourse, from social identities, from conscious and unconscious desires circulating in the community.

Understanding governance paths thus helps to discern forms of limitation and forms of self-limitation. It further helps to discern the difference between limitation and self-limitation and where either of them becomes a problem (see Chapter 11 for the implications). Awareness of limitations can enable reflection on their genesis and functioning in governance paths, while opening the door to the building of strategies that can overcome some of them (Duit & Galaz, 2008; Rasche & Seidl, 2020). Such strategies must be inspired by an understanding of the same governance path, as the reasons for the limits might still be at work, as some legacies are harder to dismantle than others, and as, simply, one must work with the materials at hand (See Chapter 10).

Self-limitation can be conscious and unconscious. When it is conscious, it can be the result of a genuine choice, or not. As genuine choice we understand a choice taken in freedom, without coercion, with all pertinent information at hand, and without anyone misrepresenting the situation. Of course, one needs to be careful here not to slide back into theories of rational decision-making and not to assume that perfect information, stable and unified reason, absence of power, or absolute self-knowledge are possible (Allmendinger, 2017; Hillier, 2003). We know from experience, from the previous chapters, and since Freud, that none of this can be assumed.

In governance, there are always reasons not to be transparent where it might have been possible, while stories, privileged forms of expertise, and the selectivities coming with the ongoing reproduction of governance, the mere application of its procedures, all contribute to a pattern of fantasies and affect, of opacity and transparency, and inclusion and exclusion that do not allow for the formation of idealized decision situations (Flyvbjerg, 2002; Forester & Fischer, 1993; Valentinov et al., 2019). Even where policy options are clearly delineated, the reasons for these and not others go back to stories, fears, and desires, to artifacts of the governance system and its cognitive and organizational capacities (Gabriel, 2015; Scherer et al., 2013). Even where one option is picked in the most transparent and even-handed manner, what makes it look desirable and real, how reality and desire entwine, and what makes it look like feasible, cannot be easily traced back. With regard to feasibility, there is always an implication of the current system of governance, its self-understanding, and its internal competition in the assessment (Clarke, 2005; Gunder & Hillier, 2004). Something looks expensive because of ideas of value, and because of the cost in current organization, and cost-benefit considerations rely on the capacity to understand benefits, including the benefits possibly accruing from actions of the governance system itself (Schiffino et al., 2017; Schlaufer, 2018). This again implies an understanding of the system and its limitations.

Limitation and self-limitation thus relate in complex manners, which might never be entirely elucidated, but where a reconstruction of governance paths is helpful for both, in providing insight and suggesting options for overcoming them. Communities tend to be unaware that governance can do more, which can mean (see also Chapter 10) that the current system could achieve more, or that a reform of the system is possible, which could extend its impact and enhance its cognitive and coordinative capacities (Jasanoff, 2020). The reasons why something is not observed as a possible way forward, or why one solution looks entirely impossible, while it is technically not, can be varied.

The knot will be unique for each community, depending on the stories about self and environment, about the tools available to change self and environment, and the contingent history of community and governance (see earlier chapters).

When limitations are understood better, and unconscious or otherwise invisible reasons for self-limitation are grappled with, self-limitation can become more freely chosen and ultimately more effective (cf. Fink, 1999). This often entails that more is possible than previously considered, in terms of higher ambitions, but also in terms of the varied paths open for exploration. It can also mean that actual limitations were not observed, leading to unobserved or belatedly noticed (financial, environmental, etc.) issues. In the envisioning of a collective future, non-reflection on limitations can give free rein to fantasies structuring desired futures that are not likely to be productive. Disappointment is likely to ensue and the response to it is risky in more than one way, as it can render future attempts at a collective strategy unpalatable, and as it can trigger resentment and other negative emotions undermining cohesion (Capelos & Demertzis, 2018; Taylor, 2009).

Self-limitation can exist under the radar for a long time. When not too much happens in the community, in the relation to its environment, the capacity to think and organize differently might not be taxed and questioned too much. If, moreover, this happens in an environment where governance is not expected to do much, or, worse, when both politics and administration are seen as a luxury, a waste, as useless, or as the enemy, then such tropes and master narratives are likely to entrench non-observation of alternatives by governance, and non-reflection on the capacity of governance to do more. Problems will appear, however, sooner or later.

Aggravation with new generations, with newcomers, with those who want something different or more diverse, and are stifled in all their actions, can stem from self-limitation (Coakes & Bishop, 1998; Cooke & Kothari, 2001). Those who think differently move out, and those who would be able to introduce new activities, lifestyles, and make the community attractive for newcomers, diversifying the economy and culture, move out. If people move in, they will tend to share or subscribe to the Master Signifiers ruling the community and its governance, with the potential of governance remaining in the shadows.

Self-limitation can create internal conflicts in the community, where it impedes governance to recognize different values and narratives existing together, or to understand how synergies could be identified between those different orientations, which might or might not coincide with different groups of people (O'Connor, 2015). Where the dominant framing narrative does not allow for the existence of alternatives alongside it, or where that or other narratives impede the actualization of the implied public goods or desired living environment, self-limitation ends up stumbling, ends up in dissatisfaction and conflict (Hamber & Wilson, 2002; Hoggett, 2006).

Many more things can go wrong when both limitation and self-limitation are not well understood, or when affective persuasion erases those understandings (Bojesen, 2018). What was seen before, can easily be ignored or forgotten, by the lure of prosperous futures, the specter of power, the desire for harmony and unity (Gunder & Hillier, 2009). Fear can be mobilized, as can restless uncertainty, by binding public discourse to new simplifications and big promises (Couperus et al., 2023). Of course, at some point the limitations of those new stories will be exposed. Ambitions will fail, ideologies will start to sound hollow, and new tautological strategies of de-paradoxification will have to be deployed by those in power (Žižek, 2012). One possible effect of an impossible drive for unity through ineffective simplification is the fragmentation of the community, or the loss of governance capacity. Where previously productive fictions might have existed, Voice disappears, both collective Ego and Superego (Van Assche, Greenwood et al., 2022).

When such problems appear, when limitation and self-limitation require more scrutiny, more self-analysis might be advisable, of the sort we called therapy earlier, to be discussed in detail in Chapter 11. Now we can safely assert that a careful reconstruction of the governance path will have to be central to such effort.

Concluding reflection: identities and boundaries

Governance can only work if it delimits itself as a system with boundaries, and its own logic of reproduction, yet the expectations of governance make it imperative that this system be utterly sensitive to signals from the outside, and all the more so in democracies. The observation of (the potential of) governance in the community can engender processes of self-organization and self-transformation that enable individuals, groups, and organizations to play a role in governance – possibly advocating for governance reform, enabling a role for the newcomers and an increase in governance capacity.

This is not merely a process of improved representation and participation; it is a creative process, whereby power emanates from two directions. In each step of transformation (in governance and in the community), it is impossible to look back objectively and assess objectively whether what happened is good, more democratic, fair and just, in reference to the old positions. Those positions have disappeared, as ideas of democracy, justice, environmental quality, the good life, fair distribution of goods, and the balance between duties and responsibilities in the community might have transformed (Van Assche, Beunen et al., 2022). If one looks back far enough, ideologies and identities have likely been reshaped. The implication is not that anything goes, but rather that *literally anything* can potentially change in governance systems and their coevolving communities, and that therefore historic points of reference, to evaluate that change, cannot be found. A current reference will have to suffice, and assessing transformation will have to be satisfied with contextual and current understandings of democracy, justice, and identity (Žižek, 1991).

What constitutes and what is understood as governance, and who is present in the community and represented in governance are redefined at the same time, with implications for the distinction between system/environment, or inside/outside for both governance system and the community itself (Frosh, 2015; Frosh et al., 2003; Valentinov et al., 2018). Reflecting on new modes of governance and new patterns of participation and representation, and inclusion and exclusion is de facto reflection on the nature of the community. What and how things are governed thus co-evolves with changes in what is imagined as governable and what is governable in practice. What is governable in practice, what is amenable to coordination, cannot be disconnected from what is imagined to be governable, or what is desired to be governable (Hinchman & Hinchman, 1997; Murdoch & Abram, 1998).

What appears as authentic, as interior, is never entirely so, and following Lacan, the folding of exteriority into identities at all levels has to be considered a driving force in the formation of unconscious drivers and limitations. As the individual, the governance system, and the community are environments (exteriorities) for each other, these processes of folding render the knot more complex, that is, the knot of the conscious and the unconscious, the pattern hence, of centering and decentering the underlying decision-making of individuals and communities (Berger, 1988).

We return to the unique position of governance, in representing a community and shaping it, in distributing and exercising power, in absorbing and creating discourse, affect, and legitimacy. Legitimacy itself can be analyzed as a hybrid of discourse and affect, as resting on formal and informal institutions, embodying a fit between governance system and environment (Scherer et al., 2013). Governance tools can become symbols of legitimacy, become invested with affect, become linked to tradition, memory, and identity, while shocks and conflicts can alter these connections, and render a tool, a concept, a narrative much less persuasive.

References

Allmendinger, P. (2017). *Planning Theory* (3rd ed., p. 346). Macmillan Education UK. https://books.google.ca/books?id=XQAoDwAAQBAJ

Benwell, B. (2006). *Discourse and Identity*. Edinburgh University Press.

Berger, B. M. (1988). Disenchanting the concept of community. *Society, 25*(6), 50–52. https://doi.org/10.1007/BF02695775

Beunen, R., Van Assche, K., & Duineveld, M. (2014). *Evolutionary Governance Theory*. Springer.

Beunen, R., Van Assche, K., & Gruezmacher, M. (2022). Evolutionary perspectives on environmental governance: Strategy and the co-construction of governance, community and environment. *Sustainabilty, 14,* 9912.

Boin, A., Ekengren, M., & Rhinard, M. (2020). Hiding in plain sight: Conceptualizing the creeping crisis. *Risk, Hazards & Crisis in Public Policy*, *11*(2), 116–138. https://doi.org/10.1002/rhc3.12193

Bojesen, E. (2018). Utopia and politics. *Policy Futures in Education*, *16*(3), 375–378. https://doi.org/10.1177/1478210317739490

Braithwaite, J. (2020). Meta governance of path dependencies: Regulation, welfare, and markets. *The ANNALS of the American Academy of Political and Social Science*, *691*(1), 30–49. https://doi.org/10.1177/0002716220949193

Brockmeier, J. (2002). Remembering and forgetting: Narrative as cultural memory. *Culture & Psychology*, *8*(1), 15–43. https://doi.org/10.1177/1354067X0281002

Capelos, T., & Demertzis, N. (2018). Political action and resentful affectivity in critical times. *Humanity & Society*, *42*(4), 410–433. https://doi.org/10.1177/0160597618802517

Clarke, J. (2005). Performing for the Public: Doubt, Desire, and the Evaluation of Public Services. In P. du Gay (Ed.), *The Values of Bureaucracy* (pp. 211–232). OUP.

Coakes, S. J., & Bishop, B. J. (1998). Where do I fit in? Factors influencing women's participation in rural communities. *Community, Work & Family*, *1*(3), 249–271. https://doi.org/10.1080/13668809808414235

Cooke, B., & Kothari, U. (2001). *Participation: The New Tyranny?* Zed books.

Couperus, S., Tortola, P. D., & Rensmann, L. (2023). Memory politics of the far right in Europe. *European Politics and Society*, *24*(4), 435–444. https://doi.org/10.1080/23745118.2022.2058757

Czarniawska, B. (2014). *A Theory of Organizing*. Edward Elgar Publishing.

Duineveld, M., Van Assche, K., & Beunen, R. (2013). Malta's unintentional consequences: Archaeological heritage and the politics of exclusion in the Netherlands. *Public Archaeology*, *12*(3), 139–154. https://doi.org/10.1179/1465518714Z.00000000039

Duineveld, M., Van Assche, K., & Beunen, R. (2017). Re-conceptualising political landscapes after the material turn: A typology of material events. *Landscape Research*, *42*(4), 375–384. https://doi.org/10.1080/01426397.2017.1290791

Duit, A., & Galaz, V. (2008). Governance and complexity—Emerging issues for governance theory. *Governance*, *21*(3), 311–335.

Elden, S. (2017). *Foucault: The Birth of Power*. John Wiley & Sons.

Fink, B. (1999). *A Clinical Introduction to Lacanian Psychoanalysis: Theory and Technique*. Harvard University Press.

Flyvbjerg, B. (2002). Bringing power to planning research. One researcher's praxis story. *Journal of Planning Education and Research*, *21*, 353–366.

Forester, J., & Fischer, F. (1993). *The Argumentative Turn in Policy Analysis and Planning*. Duke University Press.

Foucault, M. (2002a). *Archaeology of Knowledge* (2nd ed.). Routledge. https://doi.org/10.4324/9780203604168

Foucault, M. (2002b). *The Order of Things: An Archaeology of the Human Sciences*. Psychology Press.

Frosh, S. (2015). What We are Left With: Psychoanalytic Endings. In S. Frosh (Ed.), *Psychosocial Imaginaries* (pp. 200–216). Palgrave Macmillan UK. https://doi.org/10.1057/9781137388186_10

Frosh, S., Phoenix, A., & Pattman, R. (2003). Taking a stand: Using psychoanalysis to explore the positioning of subjects in discourse. *British Journal of Social Psychology*, *42*(1), 39–53. https://doi.org/10.1348/014466603763276117

Gabriel, Y. (1999). *Organizations in Depth: The Psychoanalysis of Organizations.*

Gabriel, Y. (2015). Narratives and Stories in Organizational Life. In A. D. Fina & A. Georgakopoulou (Eds.), *The Handbook of Narrative Analysis* (pp. 273–292). John Wiley & Sons, Ltd. https://doi.org/10.1002/9781118458204.ch14

Gunder, M., & Hillier, J. (2004). Conforming to the expectations of the profession: A Lacanian perspective on planning practice, norms and values. *Planning Theory & Practice*, *5*(2), 217–235. https://doi.org/10.1080/1464935041000 1691763

Gunder, M., & Hillier, J. (2009). *Planning in Ten Words or Less: A Lacanian Entanglement with Spatial Planning.* Ashgate Publishing, Ltd.

Hamber, B., & Wilson, R. A. (2002). Symbolic closure through memory, reparation and revenge in post-conflict societies. *Journal of Human Rights*, *1*(1), 35–53. https://doi.org/10.1080/14754830110111553

Hillier, J. (2002). *Shadows of Power: An Allegory of Prudence in Land-Use Planning.* Routledge.

Hillier, J. (2003). 'Agon'izing over consensus: Why Habermasian ideals cannot be 'Real'. *Planning Theory*, *2*(1), 37–59. https://doi.org/10.1177/147309520300 2001005

Hinchman, L. P., & Hinchman, S. (1997). *Memory, Identity, Community: The Idea of Narrative in the Human Sciences.* SUNY Press.

Hoggett, P. (2006). Conflict, ambivalence, and the contested purpose of public organizations. *Human Relations*, *59*(2), 175–194. https://doi.org/10.1177/00187 26706062731

Hoggett, P. (2015). *Politics, Identity and Emotion.* Routledge. https://doi.org/10.4324/9781315632704

Jasanoff, S. (2020). Imagined Worlds: The Politics of Future-Making in the Twenty-First Century. In A. Wenger, U. Jasper, & M. D. Cavelty (Eds.), *The Politics and Science of Prevision* (pp 27–44). Routledge.

Kapoor, I. (2015). What 'drives' capitalist development? *Human Geography*, *8*(3), 66–78.

Kooiman, J. (2020). Exploring the Concept of Governability. In I. Geva-May, B. G. Peters, & J. Muhleisen (Eds.), *Theory and Methods in Comparative Policy Analysis Studies.* Routledge.

Luhmann, N. (1995). *Social Systems* (Vol. 1). Stanford University Press Stanford.

Luhmann, N. (2018). *Organization and Decision.* Cambridge University Press.

Luhmann, N., Ziegert, K. A., & Kastner, F. (2004). *Law as a Social System.* Oxford University Press.

Murdoch, J., & Abram, S. (1998). Defining the limits of community governance. *Journal of Rural Studies*, *14*(1), 41–50. https://doi.org/10.1016/S0743-0167(97)00046-6

Newman, S. (2005). *Power and Politics in Poststructuralist Thought: New Theories of the Political.* Routledge. https://doi.org/10.4324/9780203015490

O'Connor, C. D. (2015). Insiders and outsiders: Social change, deviant others, and sense of community in a boomtown. *International Journal of Comparative and Applied Criminal Justice*, *39*(3), 219–238. https://doi.org/10.1080/01924036.2014.973049

Pottage, A. (1998). Power as an art of contingency: Luhmann, Deleuze, Foucault. *Economy and Society*, *27*(1), 1–27.

Rasche, A., & Seidl, D. (2020). A Luhmannian perspective on strategy: Strategy as paradox and meta-communication. *Critical Perspectives on Accounting*, *73*, 101984–101984. https://doi.org/10.1016/j.cpa.2017.03.004

Scherer, A. G., Palazzo, G., & Seidl, D. (2013). Managing legitimacy in complex and heterogeneous environments: Sustainable development in a globalized world. *Journal of Management Studies, 50*(2), 259–284. https://doi.org/10.1111/joms.12014

Schiffino, N., Taskin, L., Donis, C., & Raone, J. (2017). Post-crisis learning in public agencies: What do we learn from both actors and institutions? *Policy Studies, 38*(1), 59–75. https://doi.org/10.1080/01442872.2016.1188906

Schlaufer, C. (2018). The narrative uses of evidence. *Policy Studies Journal, 46*(1), 90–118. https://doi.org/10.1111/psj.12174

Sehring, J. (2009). Path dependencies and institutional Bricolage in Post-Soviet water governance. *Water Alternatives, 2*(1), 61–81.

Seidl, D. (2016). *Organisational Identity and Self-Transformation: An Autopoietic Perspective*. Routledge.

Simmie, J. (Ed.) (2014). Path Dependence and New Technological Path Creation in the Danish Wind Power Industry. In *Path Dependence and New Path Creation in Renewable Energy Technologies* (pp 753–772). Routledge.

Taylor, N. (2009). Tensions and contradictions left and right: The predictable disappointments of planning under new labour in historical perspective. *Planning Practice & Research, 24*(1), 57–70. https://doi.org/10.1080/02697450902742155

Valentinov, V., Roth, S., & Will, M. G. (2018). Stakeholder theory: A Luhmannian perspective. *Administration & Society, 51*(5), 826–849. https://doi.org/10.1177/0095399718789076

Valentinov, V., Verschraegen, G., & Van Assche, K. (2019). The limits of transparency: A systems theory view. *Systems Research and Behavioral Science, 36*(3), 289–300. https://doi.org/10.1002/sres.2591

Van Assche, K., Beunen, R., & Duineveld, M. (2013). *Evolutionary Governance Theory: An Introduction*. Springer.

Van Assche, K., Beunen, R., Duineveld, M., & Gruezmacher, M. (2017). Power/knowledge and natural resource management: Foucaultian foundations in the analysis of adaptive governance. *Journal of Environmental Policy & Planning, 19*, 308–322.

Van Assche, K., Beunen, R., & Gruezmacher, M. (2024). *Strategy for Sustainability Transitions: Governance, Community and Environment*. Edward Elgar Publishing.

Van Assche, K., Beunen, R., Gruezmacher, M., & Duineveld, M. (2020). Rethinking strategy in environmental governance. *Journal of Environmental Policy & Planning, 22*(5), 695–708. https://doi.org/10.1080/1523908X.2020.1768834

Van Assche, K., Beunen, R., Verweij, S., Evans, J., & Gruezmacher, M. (2022). "No time for nonsense!": The organization of learning and its limits in evolving governance. *Administration & Society, 54*(7), 1211–1225. https://doi.org/10.1177/0095399721093695

Van Assche, K., Duineveld, M., Beunen, R., Valentinov, V., & Gruezmacher, M. (2022). Material dependencies: Hidden underpinnings of sustainability transitions. *Journal of Environmental Policy & Planning, 24*(3), 281–296. https://doi.org/10.1080/1523908X.2022.2049715

Van Assche, K., Greenwood, R., & Gruezmacher, M. (2022). The local paradox in grand policy schemes. Lessons from Newfoundland and Labrador. *Scandinavian Journal of Management, 38*(3), 101212. https://doi.org/10.1016/j.scaman.2022.101212

Van Haute, P. (2002). *Against Adaptation: Lacan's "Subversion" of the Subject*. Other Press.

Ziai, A. (2004). The ambivalence of post-development: Between reactionary populism and radical democracy. *Third World Quarterly, 25*(6), 1045–1060.

Žižek, S. (1991). Formal democracy and its discontents. *American Imago, 48*(2), 181–198.

Žižek, S. (2012). *The Year of Dreaming Dangerously*. Verso Books.

9 Power, drives, and drivers in governance

Drivers of governance and Master Signifiers

Governance transforms itself through a variety of mechanisms. This set of tools will be unique and mark the identity of a governance path (Pierre & Peters, 2020; Van Assche et al., 2013). Indeed, what makes a governance system unique is not just a set of features recognizable at the time being but its path, bringing to life a limited set of modes of self-transformation that are not entirely transparent. Governance systems also rely on a particular set of couplings with the social and the ecological environments, couplings that constrain and enable self-transformation, again in ways not entirely visible to outsiders or insiders (Beunen et al., 2022; Beunen & Van Assche, 2021).

Actors in governance take decisions, and people in the community show what they want through participation, representation, and public discourse. Academics write reports for political and administrative actors, and float their ideas in papers and books, and on social and traditional media. Lobby groups can represent economic sectors or groups of organizations, while other organizations lobby for themselves. Events happen that seemingly force decisions, such as disasters, wars, economic shocks, and waves of migration, with all of these testing the infrastructures of welfare states and creating new narrative fissures.

The drivers of governance are thus, at first sight, easy enough to post. We would, however, like to return to some of the key functions and features of governance, to show that there is always more going on. Governance needs to envision a collective, come to decisions binding that collective, and then implement those decisions. Governance represents but also shapes that collective, which, in imaginary form, we called community. It can make a community more real, by doing things in line with the narratives the community tells itself, or that are told by governance about the community. This is not necessarily a matter of manipulation or of abuse of the powers of governance; images of unity and identity are needed to underpin those decisions, to unfold the paradox (Shapiro, 2006; Sundgren & Styhre, 2006).

This automatically gives a greater significance to Master Signifiers, to their structuring of key narratives and their functioning in the community.

DOI: 10.4324/9781032696676-9

A governance path can enshrine Master Signifiers while veiling them from the community, meaning that they can be at work through old encodings in institutions, or in seemingly disconnected narratives that were nevertheless structured by them. One can think of discussions on structural racism, on blind spots, on inadvertent yet persistent exclusions, or, positively, on religious values encoded in policy and whole governance systems without too many people being aware of it (Frosh, 2013; Hook, 2018). We are familiar already with Master Signifiers that are visible and openly embraced, and others that perhaps operate in the unconscious. We know from our discussion of narrative that some Master Signifiers are loosely coupled, and do not much affect each other, while others are tightly coupled and woven into the fabrics of ideology (Stavrakakis, 2008; Zaretsky, 2008).

Master Signifiers can feature in high rhetoric in governance, or they can be latent, yet intentionally cultivated (Glenn, 2004). They can be unobservable for the actors in governance and the system itself – with administrative actors assisting in the creation of a unified self-observation, an encoding of blind spots that makes them hard to dislodge. Difficulties in the observation of blind spots and latent Master Signifiers can be the result of forgetting, of repression, and trauma, as we discussed in the chapter on memory and legacy. The push to the unconscious, can also, however, and here we follow Lacan again, be the result of the silent workings of governance itself (Stavrakakis, 2002; Žižek, 2009). As the mechanics of language and signification installs itself in the human mind and reconfigures the mind and its socialization, the mechanics of governance always produces a divided subject of the community. What is invisible is not only the world of repressed ideas and affects, but also the actual logic of signification and organization that supports the operations of governance. The unique knot of organizing and thinking that supports a community and bestows upon it an identity which is not of the order of the imaginary, and is never fully accessible to the community. The need for productive fictions is the need for imaginary intervention, to bridge the divide, to make a community recognize a governance system and its decisions as its own.

The collective conscious and unconscious both feature Master Signifiers, which do not necessarily cohere in the way imaginary self-identifications suggest (Gunder & Hillier, 2016). They fulfill several functions, in suggesting cohesion, carrying and amplifying affect, and rendering situations meaningful and understandable. Their openness allows them to provide structure while inviting a variety of interpretations and affective investments. Hidden Master Signifiers can drive governance, can bestow and undermine power. They can contradict other values and aspirations, marginalize people and perspectives that are supposed to be respected, and undermine rules and values which are supposedly enshrined (Clarke, 2020; Ziai, 2016).

Master Signifiers can absorb others, override others, and concentrate affect and desire to such an extent that it looks OK to break all other rules, to sacrifice other values, to harm the community itself. Getting rid of a social ill or an evil enemy, or pursuing prosperity or equality obsessively can override and crowd out other considerations (Flescher, 1949). We refer back to our discussion of the collective Superego, which cannot be understood as representative of an agreed-upon set of values and goals, as a reasonable enforcer attuned to the affective and cognitive realities of the community. No, the Superego, armed with affectively laden Master Signifiers, can mobilize the powers of governance, and the complicity of the community, to harm governance and community itself (McGowan, 2019). The enforcer can break loose of rules itself, can lose itself in blind enforcement. This can destabilize the discursive and affective framework of community.

The Lacanian Big Other, as the symbolic order in its coercive manifestation, needs to be distinguished from the Superego, but forms a framework which co-constitutes it and from which it can be unleashed (Hook & Vanheule, 2016). Without a symbolic order bestowed with authority, no Superego can form. Without Master Signifiers, capable of attracting discourse and affect, and blinding one to differences, that Superego cannot function, and cannot be unleashed. The Big Other, we noted before, associates with governance, as governance itself comes to symbolize authority, even when it does not function authoritatively, and as the language of authority, over time, became the language of governance – as distinct from the language of brute power and individual leaders (Foucault, 2012; Salter & Dutta, 2023). Governance systems are assisted in their emergence as Big Other by their opacity and complexity, rendering it easier to project all-knowing and unified agency in it.

Drives and desires

The Superego is not only destructive because it can overlook rules and morality, because it is allowed to do so, but mostly because it connects to drives (Freud, 1989, 2003). Drives do not pursue anything, good or bad, the collective good, or disavowed desire. They repeat, have no inherent break or limitation, and are not interested in their environment. Following both Freud and Lacan, we understand drives as always there, and, even in their atavistic return to the same state, not determined by biology, or by the deep history of humanity. Indeed, they exist, but they are never disconnected from the social, and, hence, from the world of signification and its contingent histories (Phillips, 2024).

While desire might more directly connect to stories, to what appears as desirable in them, and from there, to the possibility of collective desires, drive manifests itself differently in different contexts. This might be interpreted as mere dilution of the same essence, but we prefer to see it as a transformation of drives, by the powers of the individual and the collective. We would even argue that drives can be triggered, can be materialized, through the existence of collectives and collective affects (Johnston, 2005). Understood that way, collective drives are potentialities, emerging from the history of communities, making them more predisposed to certain appearances of drives under particular conditions (Heath-Kelly, 2019).

The drivers of governance cannot be located solely in the individual or collective drives, and psychoanalysis does not suggest that the deepest, and most authentic motivations in governance can be found in unacknowledged primal communal instincts (Freud, 2004, 2015). We do believe that drives cannot be omitted from a psychoanalytically inspired perspective on governance. Collective drives emerge in the same contingent path of governance evolution that enables the appearance of the collective Ego, subject, Superego, and Big Other. Drives can appear through the discursive and affective matrix generalized through governance and can go beyond its initial boundaries because of the powers of governance (Žižek, 2009).

What looks desirable for a community can give way to the unrepentant repetition of the drive when reflection stops, when fantasy takes over the collective capacity to shift perspective, when governance can reshape the community to such an extent that stopping the repetition, and opening new avenues of reflection becomes virtually impossible. Authoritarian states that can institutionalize an ideology beyond the person of the leader, that can block opposition and alternative perspectives, can get caught in a transgressive enjoyment of utopian, aggressive, or purifying drives spiraling out of control (Žižek, 2002, 2011). The Stalinist purges reinforced themselves, and attachment to the purging itself took over an identification with the goal of the purge.

Stalinist figures and regimes are not needed, however, to trigger a mobilization of drives, as in the appearance of what Mattias Desmet calls mass formation, or collectives immunizing themselves for alternative understandings of the world, and for harm done by their own unfolding (Desmet, 2022). Mass formation, often catalyzed through gaining influence over governance, can reinforce itself in an iterative process that gives space to the drives to materialize. Shocks, and cracks in the building of reality and desire in a community open spaces for new and usually simplifying processes of mass formation, which can then unleash the power of the drives. Mass formation, but also less pernicious forms of Ego formation, can lead to a recognition of new desires, or, forms of enjoyment, while, conversely, recognition of the community of new desires, can reshape the political and cultural landscape by forming new identities (Gunder, 2003; Kinnvall & Svensson, 2022). The recognition of new desires and the formation of new identities tend to involve the mobilization of historicizing narratives. One can speak of historicizing modes of de-paradoxification, as the newly created identity is posited as always existing, and as the new desire is reinterpreted as having been always ignored, thus lending it the power of the underdog – the promised enjoyment of justice after undeserved suffering (Polletta, 1998; Renström et al., 2023).

Drives need the implication of a public. They can start from governance and hop to the community, or the other way around, yet they need the involvement of a group of people willing to stop thinking and align behind a banner giving them permission to enjoy a possibly destructive repetition of moves (McGowan, 2019; Post, 1986). Because public discourse can reflect and create a public, that is, a link between community and governance and a medium for collective affect, it will be mobilized (Carpentier et al., 2003; Wahl-Jorgensen, 2019). Brute coercion by a regime, by governance, might be effective for a while and might reflect a mobilization of drives internally, but will not last long if it does not produce a collective willing to mobilize its own drives. This, we know, does not need to coincide with the intentions and the drive of those who tried to trigger the event (Ruti, 2010, 2012). Masses, once unleashed, can go awry on a path of their own making, although that process does not need to involve the expression of previously held values and beliefs. They can be swept up, and narratives can be the excuse, the product, the veil for the drives at work (Frosh et al., 2003; Polletta, 2018).

Affect, possibly mediated through governance, can trigger the unlocking of drives, and shape the pathway of their unfolding (Vanier, 2010). Once set in action, drives can trigger new emotions, contain or override others, and transpose some into different keys. Hope and fear, aspiration and avoidance, love and hate can move to the background, or they can be unlocked by drives. Harmonizing and dividing, desiring and aggressive drives mark the functioning of governance (Gunder & Hillier, 2009; Stavrakakis, 2002). They are also projected onto governance, and as governance is expected to validate the reality of collective affects, they can reshape external reality in response to them. The promise of reality-shaping and reality-confirmation associated with governance, prepares the playing field for drives, in and through governance (Glynos & Stavrakakis, 2008).

Governance functions through the positioning of actors in fields of distinctions in discourse and affect, stipulating and creating differences that separate identities, alternative futures, and diverging interpretations of the past (Epstein, 2011; Vanheule, 2016). Governance thus leans on the production of difference, while it also requires the maintenance of unity, and this balancing act is by definition imperfect (Luhmann, 1995a). For us, the balancing act will be different for each moment and each configuration, while the imperfection leaves space for unexpected affect and the releasing of drives to occur at any moment (Van Assche et al., 2024).

To understand the role of drives as drivers in governance, it is helpful to envision the diversity of drives and the different modes of containment that, in the end work. They have to work, as otherwise the community falls apart (Kapoor, 2015; Pohl & Swyngedouw, 2023). Institutional restructuring and reinvigoration of concepts such as nation, nation-state, culture, ethnicity, and race, can bring about a reconstruction of governance and community which allows people to pretend that a community always existed, and is unproblematically self-similar, while in reality a new society is forming, using some older names, stories, and organizational forms (Frosh, 2013). Such does not contradict a co-evolutionary perspective, as also for such governance theories (such as evolutionary governance theory; Van Assche et al, 2013, 2024), shocks remain possible, and the rebuilding of societies and governance systems out of the shards of older versions allows for more radical innovation and contingency (Van Assche, Greenwood et al., 2022). An entirely different matter, however, is whether communities desire such a mode of transformation. A destructive release of drives thus promises rebirth and radical innovation, but at a great cost, which cannot be mapped out in advance (Ruti, 2012; Žižek, 2009).

If we follow Freud in distinguishing death and life drives, then we can diversify the species from there and recognize versions that are more likely than others to appear in one governance path and its community. Life drives do not aim at pleasure alone, nor at finding a balance with an environment; rather, they aim at growth, to expansion, and this led to the emergence of a reality principle in otherwise pleasure-seeking creatures (Freud, 2003). This internal split immediately leads to a qualification of drives, beyond blind repetition, and to a contextualization of drives, by a reality image created to allow for the continued seeking of pleasure while continuing to exist (Iversen, 2007). Tensions between pleasure and (the image of) reality will persist, but the reality principle inserts a reflexivity into the drive that binds it, yet also has the potential to make it stronger, by delineating a playing field in which it is likely to continue longer.

A similar internal split cannot be found with the death drives. Very different manifestations remain possible, however (Moyaert, 1996, 2007). A drive toward destruction or self-destruction, as for example, self-punishment for (collective) guilt, a drive toward ambition levels that are easy enough to recognized as unrealistic, can be considered death drives (Cucharo, 2024). Yet, also an entirely utopian wish for consensus, for a balance with the environment, can conceal death drives as much as life drives (Bigda-Peyton, 2004). A wish to avoid conflict at all cost can mask a wish not to tackle difficult decisions, a wish not to govern at all, a wish to cease existing (Dick & McLaughlan, 2020). This can be an individual desire, and it can reflect collective sentiments, possibly connected to the death drive. The cultivation of warlike virtues can represent a genuine morality of strength and discipline, and somewhere else it can be associated with a lack of interest in life itself. As always, the same practice, the same idea, and the same affect can be overdetermined in a manner that can only be reconstructed slowly, a posteriori.

Power/knowledge, narrative, and affect

Governance creates realities, and it reflects realities. It produces difference and absorbs difference, but is never able to eradicate difference, within the community and between governance and community. In the definition and pursuit of public goods, this impossibility will be visible; hence, it plays a role in the recognition of needs, the articulation of value rankings and policy hierarchies. Tension will remain, and that tension is necessary for society to reproduce itself (Hansen, 2014).

Within governance, this can be recognized in the instability of roles, where actors never see themselves in exactly the same way as other actors or as people in the community do (Van Assche et al., 2024). Their actual role is bound to differ from what they and other people believe it to be, and what they would desire it to be. The overlapping images enable functioning in and of governance, but as the different versions of roles never completely coincide, friction, reinterpretation, strategy, and competition are always there, amounting to an instability that is productive, allowing for a wide variety of responses to changing environments.

The production of realities, we know, entails both discursive and material effects, with both types of reality effects potentially reinforcing each other, leading to a naturalization of governance and society that explains current procedures and goals (Van Assche et al., 2021). Mutually supporting narratives and their Master Signifiers can naturalize a reality that seeks to stabilize itself, but they can also underpin an ideology which aims for change, a change explained by reference to the images of present and future produced by governance (Vaara et al., 2016). This, too, is a mode of de-paradoxification, since accepting that one operates only on the basis of self-constructed images would jeopardize the functioning of governance (Luhmann, 1995b).

The functioning of power in governance thus understood goes beyond the application of resources or coercion, beyond the encouragement of blind rule-following and role-taking. Power here looks very much as Foucault observed it as well, as a fine-grained texture of virtualities, which shifts by considering them, and through every application (Foucault, 2003, 2012). Power already emerges with the promise or perception of power, and the anticipation of power by others; it cannot be separated from its potential use by strategizing humans and from a context where strategy makes sense (Carter et al., 2010). Money as such, or knowledge as such, does not guarantee power, is not transferable, as money and knowledge have different effects in different systems, and depend on knowledge of the system itself (Foucault, 2002). Foucault spoke of gymnastics, and elsewhere highlighted the micro-physics of power as a game of movement and counter-movement, where changing one position in governance would produce a web of effects, according to the existing power/knowledge configurations (Jessop, 2007).

The virtuality of power is especially stringent in governance, as governance operates on *the promise of power*, as much as on the exercise of power. Actors want to be an actor because of that promise and compete for impact over rule-making once in governance. Those who cannot be described as actors, might still have opinions and feelings that can be mobilized by others, as part of their strategy to participate in governance. The lure of power is always there, and triggers movements in society and in governance – tactics and strategies, actions and stories, modes of organization and disorganization, initiated by those interested in power (McGowan, 2019; Newman, 2005). For Foucault, intention is not a necessity, since *anything* in society can potentially be mobilized by others interested in power further on. A lack of intention on all sides and at all times still allows for shifts in the matrix of virtualities that can be grasped by those who have to do something in governance (Shields, 2005, 2012). This can be the taking of important decisions, but also the routines of small administrative roles. It can affect the stories told at the highest leadership tables, and the route taken by the man with the coffee cart, the sequence in which the latest gossip is heard, the mood set at the meeting right after the coffee break (Åkerström et al., 2021; Bissell, 2010).

As governance can generate reality effects, and as people know this, knowledge of the governance system and the knowledges in the system acquire immediate strategic importance. If narratives become incorporated into the system through codification into institutions, through embedding in ideologies or organizational cultures; if forms of expertise, as in perspectives, methods, or individual concepts become anchored into procedures associated with important topics of decision-making, this makes them immediately influential, even if those involved are not aware of it. It makes these narratives immediately attractive for those who can see possibilities of use and abuse, and immediately attractive to critique and doubt. Any story or any idea finding itself, strategically or coincidentally, in the governance system will most likely encounter competition, even if those introducing it had no idea that such competition existed or would be necessary.

Power/knowledge permeates governance in every fiber of its fabric, and shifts in power/knowledge have the tendency to resonate far from the initial site (Latour, 2004). If departments see that a more quantitative approach to an issue leads management and politicians to bend more easily to their recommendations, this can inspire other departments to follow a similar approach, and politicians to embrace to shield themselves from public debate, while in the longer term the success of the method can change the definition of problems amenable to administrative analysis, and from there, the range of topics deemed amenable to governance (Alvesson & Spicer, 2010, 2016). If democracy comes to be associated with transparency, and transparency with accounting, then one can expect a proliferation of accounting techniques, which, in the end create opportunities for manipulation, Byzantine practices of cooking the books, and blind spots by reducing the goals of governance to that which can be counted (Czarniawska, 2010; Hood & Peters, 2004).

Leadership, ethics, and politics

The drivers of governance thus look more varied than before, and the interplay of power/knowledge and the other configurations in governance systems brings us to state that it is technically impossible to map all drivers in a system, and all forces at work behind a particular decision or policy (Van Assche et al., 2019). Discursive dynamics cannot be stopped; hence, governance systems cannot be stabilized indefinitely, even if no institutions change and if all actors still agree on existing rules. Stories find new meanings, are altered themselves, produce new affects, and connect differently to desires and drives. Discursive changes turn features of the environment into assets, activities into work, capacities of people into marketable skills, and create opportunities to capitalize on them, economically or politically (Hacking, 1990; McGowan, 2023).

As a mapping of power/knowledge and a mapping of the full multiplicity of drivers of governance cannot be comprehensive, this also means that one can never fully comprehend why certain policies and plans, why certain administrative procedures and resulting decisions work, and others less so. Not seeing why a mechanism is there suggests that the effects are less likely to be clear. Of course, drivers do not fully explain effects, and here important reasons for opacity can be found in the common blindness for informality with actors and observers alike, and the reliance on unreliable self-descriptions in governance. We refer to our earlier passages on blind spots for a detailed discussion.

The embedding of governance systems in communities with which they share informal coordination mechanisms, and the formation of organizational cultures within governance that can associate with more informal institutions, add to the complexity of power/knowledge interactions in governance, and of the relations between governance and community (Dryzek, 1996; Roy, 2005). Not grasping these informalities is not seeing the full extent of reality effects and dependencies in governance. Yet, that full spectrum would not be visible anyway. A reliance on narratives of self, of good governance, of virtues and vices, comes with constraints for the observation and understanding of features that do not clearly fit or contrast (Shanahan et al., 2011).

Partial understanding of drivers and imperfect insight into effects of governance add up to an always less-than-ideal image of how governance works, what it could do, and what can be expected of existing institutions (Luhmann, 1997; Valentinov et al., 2019). Conversely, as just noted, images of governance set participants and observers alike on the path to problematic analyses of drivers and effects. We are back in the world of narratives and Master Signifiers, of self-observing identities, and of the mutual structuring of discourse and desire.

One powerful and traditional ways of managing this opacity is leadership, the concentration of power in the hands of a few. Leadership can be distributed, and usually is, and it is a necessity for the functioning of each organization and certainly each governance system, as no system has a perfect knowledge of itself (Bolden et al., 2023). No set of institutions sets itself in motion and keeps producing the desired results. In governance, the judgment of leadership can decide which drivers and effects are relevant in each situation, which tool can be used to address a problem, which forms of knowledge can be produced or harnessed, which actors incentivized, and which procedures started. Leadership can silently be allowed to set the rules to break the rules, to prioritize topics, to shift between insider and outsider perspectives, and to observe internal and external informalities and find ways to capitalize on them (Alvesson & Einola, 2019; Ancarani et al., 2021).

Judgment is necessary for decisions, as, we know, decisions do not follow from knowledge or procedures. Hence, leadership has become associated with decision-making (Alvesson & Spicer, 2010) . In governance systems, those signing off are not always those who showed leadership, or those who came up with the idea, built support, found resources, or identified suitable policy tools (Flyvbjerg, 1998), but that is not necessarily a problem. As long as leadership functions appear in the system, as long as leadership is considered legitimate and its activities are in line with the goals of the system, there is no problem (Mintzberg, 1989). The actual form and functions of leadership will, however, differ according to governance path, and hinge on the relation between governance and community.

One can say, therefore, that leadership is expected to deal with the imperfections that come with any formal system of governance, with any system trying to formalize the nexus between thinking and organizing, and the link between individual and community (Czarniawska & Gagliardi, 2003). It can also be understood as an impossible task, certainly if bestowed upon an individual, but even if open channels of talent development and diverse forms of public participation allow for a flexible leadership group or network. As one might suspect, paradox and the need for de-paradoxification are lurking in the shadows again (Luscher, 2018; Seidl, 2016). The desire for leadership is an impossible desire because it implies the possibility of perfect knowledge of governance and environment, of an amazing set of skills to tell the truth and hide it, to tell stories necessary for the art of governing, as well as true stories. Leadership can only function as a corrective for imperfect governance if it is perfect itself, if it is clear-eyed about the functioning of governance in its environments, if it can see everything that is true and hide some of what is true for those who must be persuaded but do not want to know. Leadership is expected to hide the fact that de-paradoxification is operating all the time (Luhmann, 2018). Pretending that the procedures of governance and the political programs that are supposed to steer them are indeed coherent and effective, is part of the job, as well as keeping up the appearance that policies represent an objectively correct response of a clearly delineated identity to neutrally observed problems and opportunities. Neither overcoming the imperfections of governance nor hiding them perfectly is humanly possible, and for now at least, leaders tend to be humans.

Leadership, history shows, further represents the paradox that concentrating power is effective to build systems and to undermine systems. Distributing power over a fine-grained maze of actors makes it harder to steer that maze in a desirable direction, and it also makes it less accountable (Blair, 2000; Cooke & Kothari, 2001). While concentrating power in the hands of a few, accepting the associated efficiency and political astuteness that also come with a degree of opacity, assumes a similar risk (Mansfield, 1998). Furthermore, leadership is structurally tied to ethical dilemmas, going back to the tension between the function of governance to represent existing identities and aspirations and its potential, often desired in the community, to reshape the community, to bring to life new identities by speaking to them as if they already existed.

When shocks occur, and leadership is all the more needed, when binding narratives are sorely tested, and reality effects of governance are temporarily weakened, such ethical dilemmas come to the foreground (Boin, 2009; Boin et al., 2013). Hence the common practice to vote wartime leaders out after the war, as the wartime community does not exist anymore, as the community does not want to be confronted anymore with the need it felt for, the identification it felt with, forms of leadership that look and feel constraining and infantilizing now (Glad, 2002; Pusca, 2007). Even without a war, shocks call for leadership as routine governance does not work anymore, as temporary measures are called for measures that, de facto, suspend the current rulebook, and bestow a measure of power on the leader which cannot be fully comprehended, because of the shock and its effects (Eisenstein & McGowan, 2012; Van Assche, Gruezmacher et al., 2022). Where no shock at all occurred, calls for change or for dramatically raised ambitions will still come with a desire for and an acceptance of new forms of leadership (Collinson, 2011).

Both the imposed change of a shock, and the desire for radical change, associated with positive or negative affect, can thus open new possibilities for leadership, new roles, and new modes of concentrating power through the promise of better a future. Master Signifiers that emerge as central under such conditions tend either to entrench or to reject older master signifiers rather than open a diversity of futures (Davidson, 2010; Žižek, 2008). In charismatic leadership, the person of the leader himself or herself is presented as the guarantee for the desired future, independent of existing ideologies (Antonakis, 2012). In systems that can maintain pluralism, where the stories of legitimacy hold steady, the system itself can be seen as a guarantee for its own transformation (McSwite, 1997). In that case, the web of existing narratives can be the starting point of change. Less common is the appearance of leadership that can explore a wider variety of possible and desirable futures with the community and guide the community in that desired direction, after one is decided upon collectively (Johnston, 2009; Ruti, 2012) (see also Chapter 10).

Where ambitions and desire for change are not that outspoken, public expectations might not attract aspiring leadership figures and bring functions into being that concentrate power and skirt checks and balances for the greater good in dramatic circumstances. Such lower pressure on the maintenance of democratic values does not, however, in any sense guarantee that democracy will maintain itself. There is no logical necessity for leadership functions to appear that will compensate for the imperfections of the formal system, functions enhancing adaptive or strategic capacity (Valentinov, 2014). Another ethical dilemma asserts itself in that it is impossible *not* to take seriously the word of (aspiring) leadership regarding the flaws of the system and its vision for improving it, while taking it at face value is equally difficult. The issues are important enough to take seriously, and insider knowledge is required, while too much knowledge can associate with responsibility and guilt. Insider perspectives are immediately associated with the observed flaws, and their discourse sidelined as riddled with blind spots. A pressure to differentiate, to find arguments for another term in power (beyond housekeeping) or for coming into power, is always there. Even maintaining the course cannot be simple repetition, except where a crisis is constructed, where there is an urgent need to cling to old values and identities (Desmet, 2022).

One way to evade dilemmas is rhetorical. Leadership discourse itself tends to be codified, allowing for several interpretations, inventing a common ground where it does not exist (A. Spicer, 2017). Nowadays, after decades of neoliberal influence on public sector thinking, often under the name of new public management (Hood & Peters, 2004), this often means leaning on management school jargon and borrowing prestige from the hegemony of economic discourse and private sector organizations (Parker, 2018). Another maneuver is fetishizing experience, and ideally private sector experience. Leadership experience (independent of success) then becomes a proof of leadership skills that can be applied anywhere, and that can be associated with vision, a vision into the future, and insight into an organization (or a governance system). We encounter the same self-referential thinking that plagues governance, but now with those who claim to expose and fix the problems of self-reference in governance (Azfar Nisar, 2020; Lezaun & Muniesa, 2017).

While leadership unavoidably comes with ethical dilemmas, and while administrative leadership always has to tread carefully in stepping over into politics, the reality of governance is that the line between politics and administration cannot be stabilized, and that new leadership *functions* emerge all the time (M. W. Spicer, 2005). This is the case because the functioning of, the demands on, and the environments of governance are changing all the time. As noted, a system can work to the contentment of most citizens, until a new ideology rears its head, or a new concept of justice, or a new feeling about an old problem, and suddenly, the system is felt as hopelessly inadequate, as always already blind, and leadership, sometimes for seemingly insignificant signifiers, can appear hopelessly outdated. A mustache, a hat, a gesture, a style of joking or camaraderie can suddenly appear as the proof of a rotten system, of a generation blinded by ideology or simple greed, leaving us with a mountain of problems we must deal with (Hellmann, 2019).

Leadership has to figure prominently in the understanding of change, in the story of drivers of governance, while leadership can only be successful if it shows an understanding of other drivers, both in a technical sense and in the senses we could uncover with the help of psychoanalysis. Leadership can mobilize affect, forge coalitions, revive identities, speak to existing desires, and trigger new ones. It can rely on new versions of the past and old versions of the future, create new inclusions and inclusions, and it can persuade the community that there is no split between governance and community, that what they really want is also what the leader wants them to be.

None of this takes place in isolation. Meta-narratives on good governance, the good community, on leadership itself, will mold the matrix of possibilities for leadership to manage the relationship between the reflective and shaping functions of governance. Ethical dilemmas face leadership, but ethics itself does not need to be questioned in the community. What counts as ethical and unethical can be understood as a distillation of values that serve to stabilize both governance and community, constraining the lure of power and installing a hope that drives can be either managed or blocked from appearing as drivers in governance (Mansfield, 1998; Žižek, 2009). The truth of ethical norms cannot be ascertained, but finding or representing truth is not their function. They can codify self-reference, by stopping reflection and positing implied values as absolute (Valentinov et al., 2019).

Sublimation can transform desire, channel drives, and take on collective forms, not just for leadership but for all involved in governance, and for those in the community entrusting governance with the actualization of noble aspirations (Gay, 1992; Moyaert, 1995). Sublimation can serve as a counterweight to the seductions of governance, including the always present lure of factionalism and corruption. Positive ideals can be felt just as real as the reality of, and often easier opportunities for self-enrichment, or of simple and dutiful fulfillment of a role (Tencati et al., 2020). More than avoiding inherent moral risks of governance, which can come at great social cost, sublimation can harness creative capacities in governance and community, turning aggressive drives to work in building ambitious new futures, transforming desire for ethnic purity into creative thinking, organizing welcoming environments, and turning the guilt about a history of resource extraction into a persistent search for sustainable development.

Concluding

Governance evolves for many reasons; the drivers are diverse and many have been identified in this and previous chapters. Psychoanalysis offers new insights into that variety, in drivers that have not been recognized in most perspectives on policy, politics, and administration (Stavrakakis, 2002). Desires and drives, unconscious Master Signifiers, collective affect, Ego, and Superego all contribute, in ways explored throughout this book, to processes of reproduction and change in governance.

In this chapter, we pieced together a picture of drivers revolving, as the rest of our story, around a concept of narrative. Master Signifiers, desire, and drivers require the medium of story to function in governance, that is, to be deployed strategically and to function latently with maximum impact. Their power does not exist by itself, as governance is always an interplay of desires, narratives, and affects with an eye on giving them a freer rein in a community reshaped by governance. Power thus comes from all sides, and following Foucault, and governance theories inspired by him (such as evolutionary governance theory), we have placed power/knowledge at the heart of governance configurations. Actors use, shape, and are shaped by narratives, and institutions assume and reinforce narratives. Expert knowledge is underlain by narrative constructs, of theoretical, ideological, literary, and other natures, and its use in governance and society at large is structured by narrative and the derived expectations (Fischer, 2000; Yanow, 2000).

Power/knowledge configurations in governance are infused with, structured by desire, and, conversely, affect the circulation of desire in the community (Gherardi, 2004). A basic psychoanalytic insight representing a significant difference with mainstream theories is that what is accepted as real, is felt as real, and what is felt as real, can be so because it is desired. In the meantime, what is not desirable or acceptable will not be observed or not seen as real. Knowledge confers power, and power generates knowledge in patterns only explainable if one takes the dialectics of cognition and desire seriously; this dialectic affects potentially anything in governance, from the construction of identities framing observation in governance and community, to the affective reception of policies whose signifiers might or might not trigger rhizomatic resonances with individuals (Pottage, 1998).

The effective functioning of governance requires leadership, a concentration of power necessary because a very fine-grained distribution of roles creates steering problems and because the best rules tend not to come out of procedures, while even the best rules do not implement themselves. We have argued that leadership is always tied together with ethical dilemmas, as governance itself has the dilemma of both representing and shaping the community, with the question always arising whether the representing is done fairly (in a process of continuous interpretation and selection) and who is asking for the shaping. The internal split in the community caused by the emergence of a steering system, of governance, cannot be undone, and starts a path of mirroring, of projection of desire, of fear and hope. It will be a history of disappointment, given the non-identity of community and governance, the development of internal organizational identities, of an internal logic of reproduction, of aspirations for power itself in governance. Moreover, impossible expectations of life, of the perfect community, are most likely projected onto governance, without understanding the functioning and limitations of governance, and without governance and its leadership asking the community to reflect on its own desires and expectations (Mannheim, 2013; Žižek, 2006, 2009).

What Chapter 11, our final chapter, will argue is that some circumstances do ask for more courageous leadership and more honest citizenship, for enhanced processes of self-reflection that could interrupt the imaginary cycles of alienation between governance and community, and the inclination to find refuge in impossible or unjust futures. What psychoanalysis does is intimate more generally that both the construction and embrace of stories in governance betrays a variety of desires, needs, and drives at work, some clear to the community, some less so. The existence of governance configurations and of specialized roles in administration, the presence of checks and balances, can all contribute, as a significant side effect of their functioning, to the clarification of this variety.

It bears repeating that governance creates and absorbs difference and associated affects; it copies, modifies, reflects, and creates discourse, thus rendering it hard to demarcate the community identity without reference to the system supposed to represent that identity. Simply pointing out the policy consequences of an identity that is supposed to be out there, visible and relevant to all, existing independently of governance, of shifting stories and desires, is rarely possible. Governance shapes identity, and offers the potential of both unity and fragmentation, of stability and instability, of rigid certainty and fluid adaptation. The nature of discourse and the nature of governance entwine here to generate the potential for both construction and destruction, the potential to construct a wide diversity of forms of organizing and thinking that can both guide and reflect a community.

References

Åkerström, M., Jacobsson, K., Andersson Cederholm, E., & Wästerfors, D. (2021). *Hidden Attractions of Administration: The Peculiar Appeal of Meetings and Documents*. Taylor & Francis. https://doi.org/10.4324/9781003108436

Alvesson, M., & Einola, K. (2019). Warning for excessive positivity: Authentic leadership and other traps in leadership studies. *The Leadership Quarterly*, *30*(4), 383–395. https://doi.org/10.1016/j.leaqua.2019.04.001

Alvesson, M., & Spicer, A. (2010). *Metaphors We Lead By: Understanding Leadership in the Real World*. Routledge.

Alvesson, M., & Spicer, A. (2016). *The Stupidity Paradox: The Power and Pitfalls of Functional Stupidity at Work*. Profile Books.

Ancarani, A., Arcidiacono, F., Mauro, C. D., & Giammanco, M. D. (2021). Promoting work engagement in public administrations: The role of middle managers' leadership. *Public Management Review*, *23*(8), 1234–1263. https://doi.org/10.1080/14719037.2020.1763072

Antonakis, J. (2012). Transformational and Charismatic Leadership. In *The Nature of Leadership* (pp. 256–288).

Azfar Nisar, M. (2020). Practitioner as the Imaginary Father of Public Administration: A psychoanalytic critique. *Administrative Theory & Praxis*, *42*(1), 44–61. https://doi.org/10.1080/10841806.2019.1589230

Beunen, R., & Van Assche, K. (2021). Steering in governance: Evolutionary perspectives. *Politics and Governance*, *9*(2), 365–368. https://doi.org/10.17645/pag.v9i2.4489

Beunen, R., Van Assche, K., & Gruezmacher, M. (2022). Evolutionary perspectives on environmental governance: Strategy and the co-construction of governance, community and environment. *Sustainabilty*, *14*, 9912.

Bigda-Peyton, F. (2004). When drives are dangerous: Drive theory and resource overconsumption. *Modern Psychoanalysis*, *29*(2), 251.

Bissell, D. (2010). Passenger mobilities: Affective atmospheres and the sociality of public transport. *Environment and Planning D: Society and Space*, *28*(2), 270–289. https://doi.org/10.1068/d3909

Blair, H. (2000). Participation and accountability at the periphery: Democratic local governance in six countries. *World Development*, *28*(1), 21–39. https://doi.org/10.1016/S0305-750X(99)00109-6

Boin, A. (2009). 13. Crisis Leadership in Terra Incognita: Why Meaning Making is Not Enough. In P. t' Hart & K. Tindall (Eds.), *Framing the Global Economic Downturn* (pp. 309–314). ANU Publisher.

Boin, A., Kuipers, S., & Overdijk, W. (2013). Leadership in Times of Crisis: A Framework for Assessment. In P. `t Hart & K. Tindall (Eds.), *Leadership in Times of Crisis: A Framework for Assessment.* (Vol. 18, pp. 79–91). ANU Press. http://doi.org/10.22459/FGED.09.2009

Bolden, R., Gosling, J., & Hawkins, B. (2023). *Exploring Leadership: Individual, Organizational, and Societal Perspectives.* Oxford University Press.

Carpentier, N., Rico, L., & Servaes, J. (2003). Community media: Muting the democratic media discourse? *Continuum, 17,* 51–68. https://doi.org/10.1080/1030431022000049010

Carter, C., Clegg, S., & Kornberger, M. (2010). Re-framing strategy: Power, politics and accounting. *Accounting, Auditing & Accountability Journal, 23*(5), 573–594. https://doi.org/10.1108/09513571011054891

Clarke, M. (2020). Eyes wide shut: The fantasies and disavowals of education policy. *Journal of Education Policy, 35*(2), 151–167. https://doi.org/10.1080/02680939.2018.1544665

Collinson, D. (2011). Critical Leadership Studies. In A. Bryman, K. Grint, & D. L. Collinson, (Eds.), *The Sage Handbook of Leadership* (pp. 181–194). SAGE.

Cooke, B., & Kothari, U. (2001). *Participation: The New Tyranny?* Zed books.

Cucharo, S. J. (2024). *Guilty Subjects, Reparative Politics: On Guilt and Political Theory after Freud* [UCLA], PhD dissertation. https://escholarship.org/uc/item/99h7k3n2

Czarniawska, B. (2010). Translation impossible? Accounting for a city project. *Accounting, Auditing & Accountability Journal, 23*(3), 420–437. https://doi.org/10.1108/09513571011034361

Czarniawska, B., & Gagliardi, P. (2003). *Narratives We Organize By* (Vol. 11). John Benjamins Publishing.

Davidson, D. J. (2010). The applicability of the concept of resilience to social systems: Some sources of optimism and nagging doubts. *Society & Natural Resources, 23*(12), 1135–1149. https://doi.org/10.1080/08941921003652940

Desmet, M. (2022). *The Psychology of Totalitarianism.* Chelsea Green Publishing.

Dick, M.-D., & McLaughlan, R. (2020). Conclusion: Death Drive Ecologies. In M.-D. Dick & R. McLaughlan (Eds.), *Late Capitalist Freud in Literary, Cultural, and Political Theory* (pp. 157–174). Springer International Publishing. https://doi.org/10.1007/978-3-030-47194-1_6

Dryzek, J. S. (1996). The informal Logic of Institutional Design. In R. E. Goodin (Ed.), *The Theory of Institutional Design* (pp. 103–125). Cambridge University Press Cambridge.

Eisenstein, P., & McGowan, T. (2012). *Rupture: On the Emergence of the Political.* Northwestern University Press.

Epstein, C. (2011). Who speaks? Discourse, the subject and the study of identity in international politics. *European Journal of International Relations, 17*(2), 327–350. https://doi.org/10.1177/1354066109350055

Fischer, F. (2000). *Citizens, Experts, and the Environment: The Politics of Local Knowledge.* Duke University Press.

Flescher, J. (1949). Political life and super-ego regression. *The Psychoanalytic Review (1913-1957), 36*(4), 416.

Flyvbjerg, B. (1998). *Rationality and Power: Democracy in Practice.* University of Chicago press.

Foucault, M. (2002). *Archaeology of Knowledge* (2nd ed.). Routledge. https://doi.org/ 10.4324/9780203604168

Foucault, M. (2003). *Society Must Be Defended: Lectures at the College de France, 1975-76.* Allen Lane The Penguin Press.

Foucault, M. (2012). *Discipline and Punish: The Birth of the Prison.* Knopf Doubleday Publishing Group.

Freud, S. (1989). *The Ego and the Id* (J. Strachey, Ed.). Norton.

Freud, S. (2003). *The Uncanny* (H. Haughton, Ed.; D. McLintock, Trans.; Illustrated edition). Penguin Classics.

Freud, S. (2004). *Totem and Taboo* (2nd ed.). Routledge. https://doi.org/10.4324/ 9780203164709

Freud, S. (2015). *Civilization and Its Discontents.* Broadview Press.

Frosh, S. (2013). Psychoanalysis, colonialism, racism. *Journal of Theoretical and Philosophical Psychology, 33*(3), 141–154. https://doi.org/10.1037/a0033398

Frosh, S., Phoenix, A., & Pattman, R. (2003). Taking a stand: Using psychoanalysis to explore the positioning of subjects in discourse. *British Journal of Social Psychology, 42*(1), 39–53. https://doi.org/10.1348/014466603763276117

Gay, V. P. (1992). *Freud on Sublimation: Reconsiderations.* State University of New York Press.

Gherardi, S. (2004). Knowing as Desire: Dante's Ulysses at the End of the Known World. In Y. Gabriel (Ed.), *Myths, Stories and Organization: Premodern Narratives for our Times* (pp 174–194). Oxford University Press.

Glad, B. (2002). Why tyrants go too far: Malignant narcissism and absolute power. *Political Psychology, 23*(1), 1–2. https://doi.org/10.1111/0162-895X.00268

Glenn, P. F. (2004). The politics of truth: Power in Nietzsche's epistemology. *Political Research Quarterly, 57*(4), 575–583. https://doi.org/10.1177/10659129040 5700406

Glynos, J., & Stavrakakis, Y. (2008). Lacan and political subjectivity: Fantasy and enjoyment in psychoanalysis and political theory. *Subjectivity, 24*(1), 256–274. https://doi.org/10.1057/sub.2008.23

Gunder, M. (2003). Passionate planning for the others' desire: An agonistic response to the dark side of planning. *Progress in Planning, 60*(3), 235–319. https://doi.org/ 10.1016/s0305-9006(02)00115-0

Gunder, M., & Hillier, J. (2009). *Planning in Ten Words or Less: A Lacanian Entanglement with Spatial Planning.* Ashgate Publishing, Ltd.

Gunder, M., & Hillier, J. (2016). *Planning in Ten Words or Less: A Lacanian Entanglement with Spatial Planning.* Routledge.

Hacking, I. (1990). *The Taming of Chance.* Cambridge University Press.

Hansen, A. D. (2014). Laclau and Mouffe and the ontology of radical negativity. *Distinktion: Journal of Social Theory, 15*(3), 283–295. https://doi.org/10.1080/ 1600910X.2014.973895

Heath-Kelly, C. (2019). Forgetting ISIS: Enmity, drive and repetition in security discourse. In D. B. Monk (Ed.), *"Who's Afraid of ISIS?"* (pp 85–99) Routledge.

Hellmann, O. (2019). The visual politics of corruption. *Third World Quarterly, 40*(12), 2129–2152. https://doi.org/10.1080/01436597.2019.1636224

Hood, C., & Peters, G. (2004). The middle aging of new public management: Into the age of paradox? *Journal of Public Administration Research and Theory, 14*(3), 267–282.

Hook, D. (2018). Racism and jouissance: Evaluating the "racism as (the theft of) enjoyment" hypothesis. *Psychoanalysis, Culture & Society, 23*(3), 244–266. https://doi.org/10.1057/s41282-018-0106-z

Hook, D., & Vanheule, S. (2016). Revisiting the master-signifier, or, Mandela and repression. *Frontiers in Psychology, 6*. www.frontiersin.org/articles/10.3389/fpsyg.2015.02028

Iversen, M. (2007). *Beyond Pleasure: Freud, Lacan, Barthes*. Penn State Press.

Jessop, B. (2007). *State-Power. A Strategic-Relational Appraoch*. Polity Press.

Johnston, A. (2005). *Time Driven: Metapsychology and the Splitting of the Drive*. Northwestern University Press.

Johnston, A. (2009). *Badiou, Zizek, and Political Transformations: The Cadence of Change*. Northwestern University Press.

Kapoor, I. (2015). What 'drives' capitalist development? *Human Geography, 8*(3), 66–78.

Kinnvall, C., & Svensson, T. (2022). Exploring the populist 'mind': Anxiety, fantasy, and everyday populism. *The British Journal of Politics and International Relations, 24*(3), 526–542. https://doi.org/10.1177/13691481221075925

Latour, B. (2004). *Politics of Nature: How to Bring the Sciences into Democracy*. Harvard University Press. https://doi.org/10.4159/9780674039964

Lezaun, J., & Muniesa, F. (2017). Twilight in the leadership playground: Subrealism and the training of the business self. *Journal of Cultural Economy, 10*(3), 265–279. https://doi.org/10.1080/17530350.2017.1312486

Luhmann, N. (1995a). *Social Systems* (Vol. 1). Stanford University Press Stanford.

Luhmann, N. (1995b). The Paradoxy of observing systems. *Cultural Critique, 31*, 37–55. https://doi.org/10.2307/1354444

Luhmann, N. (1997). The control of intransparency. *Systems Research and Behavioral Science, 14*(6), 359–371. https://doi.org/10.1002/(SICI)1099-1743(199711/12)14:6<359::AID-SRES160>3.0.CO;2-R

Luhmann, N. (2018). *Organization and Decision*. Cambridge University Press.

Luscher, L. (2018). *Managing Leadership Paradoxes*. Routledge. https://doi.org/10.4324/9781351019941

Mannheim, K. (2013). *Ideology and Utopia*. Routledge. https://doi.org/10.4324/9781315002828

Mansfield, H. C. (1998). *Machiavelli's Virtue*. University of Chicago Press.

McGowan, T. (2019). Law and Superego. In Y. Stavrakakis (Ed.), *Routledge Handbook of Psychoanalytic Political Theory* (pp. 139–150). Routledge.

McGowan, T. (2023). Sublimating the Commodity. In H. Richter (Ed.), *Viral Critique* (pp 177–195). Routledge.

McSwite, O. C. (1997). Jacques Lacan and the theory of the human subject: How psychoanalysis can help public administration. *American Behavioral Scientist, 41*(1), 43–63. https://doi.org/10.1177/0002764297041001005

Mintzberg, H. (1989). *Mintzberg on Management: Inside Our Strange World of Organizations*. Simon and Schuster.

Moyaert, P. (1995). Ethiek En Sublimane. Over de Ethiek van de Psychoanalyse van Jacques Lacan. *Tijdschrift Voor Filosofie, 57*(2), 374–375.

Moyaert, P. (1996). Lacan on neighborly love: The relation to the thing in the other who is my neighbor. *Epoché: A Journal for the History of Philosophy*, *4*(1), 1–31. https://doi.org/10.5840/epoche1996417

Moyaert, P. (2007). Can sublimation be brought about through idealization? *Ethical Perspectives*, *14*(1), 53–78. https://doi.org/10.2143/EP.14.1.2021812

Newman, S. (2005). *Power and Politics in Poststructuralist Thought: New Theories of the Political*. Routledge. https://doi.org/10.4324/9780203015490

Parker, M. (2018). Shut Down the Business School. In *University of Chicago Press Economics Books*. https://ideas.repec.org/b/ucp/bkecon/9780745399171.html

Phillips, A. (2024). *On Giving Up*. Farrar, Straus and Giroux.

Pierre, J., & Peters, B. G. (2020). *Governance, Politics and the State*. Edward Elgar. https://public.ebookcentral.proquest.com/choice/publicfullrecord.aspx?p=6234793

Pohl, L., & Swyngedouw, E. (2023). Enjoying climate change: *Jouissance* as a political factor. *Political Geography*, *101*, 102820. https://doi.org/10.1016/j.polgeo.2022.102820

Polletta, F. (1998). Contending stories: Narrative in social movements. *Qualitative Sociology*, *21*(4), 419–446. https://doi.org/10.1023/A:1023332410633

Polletta, F. (2018). Participatory Enthusiasms: A Recent History of Citizen Engagement Initiatives. In *The Participatory Democracy Turn*. Routledge.

Post, J. M. (1986). Narcissism and the Charismatic leader-follower relationship. *Political Psychology*, *7*(4), 675–688. https://doi.org/10.2307/3791208

Pottage, A. (1998). Power as an art of contingency: Luhmann, Deleuze, Foucault. *Economy and Society*, *27*(1), 1–27.

Pusca, A. (2007). Shock, therapy, and postcommunist transitions. *Alternatives*, *32*(3), 341–360. https://doi.org/10.1177/030437540703200304

Renström, E. A., Bäck, H., & Carroll, R. (2023). Threats, emotions, and affective polarization. *Political Psychology*, *44*(6), 1337–1366. https://doi.org/10.1111/pops.12899

Roy, A. (2005). Urban informality: Towards an epistemology of planning. *Journal of the American Planning Association*, *71*(2), 147–158.

Ruti, M. (2010). *A World of Fragile Things: Psychoanalysis and the Art of Living*. State University of New York Press.

Ruti, M. (2012). *The Singularity of Being: Lacan and the Immortal Within*. Fordham University Press.

Salter, L. A., & Dutta, M. J. (2023). The algorithmic big other: Using Lacanian theory to rethink control and resistance in platform work. *Distinktion: Journal of Social Theory*, 1–16. https://doi.org/10.1080/1600910X.2023.2224521

Seidl, D. (2016). *Organisational Identity and Self-Transformation: An Autopoietic Perspective*. Routledge.

Shanahan, E. A., Jones, M. D., & McBeth, M. K. (2011). Policy narratives and policy processes. *Policy Studies Journal*, *39*(3), 535–561. https://doi.org/10.1111/j.1541-0072.2011.00420.x

Shapiro, S. (2006). The Governance of Identity. In *Identity and Modality* (pp. 164–173).

Shields, R. (2005). *The Virtual*. Routledge.

Shields, R. (2012). Cultural topology: The seven bridges of Königsburg, 1736. *Theory, Culture & Society*, *29*(4–5), 43–57. https://doi.org/10.1177/0263276412451161

Spicer, A. (2017). *Business Bullshit*. Routledge. https://doi.org/10.4324/9781315692494

Spicer, M. W. (2005). *Public Administration and the State: A Postmodern Perspective*. University of Alabama Press.

Stavrakakis, Y. (2002). *Lacan and the Political*. Routledge.

Stavrakakis, Y. (2008). Peripheral vision: Subjectivity and the organized other: Between symbolic authority and fantasmatic enjoyment. *Organization Studies, 29*(7), 1037–1059. https://doi.org/10.1177/0170840608094848

Sundgren, M., & Styhre, A. (2006). Leadership as de-paradoxification: Leading new drug development work at three pharmaceutical companies. *Leadership, 2*(1), 31–52. https://doi.org/10.1177/1742715006060652

Tencati, A., Misani, N., & Castaldo, S. (2020). A qualified account of supererogation: Toward a better conceptualization of corporate social responsibility. *Business Ethics Quarterly, 30*(2), 250–272. https://doi.org/10.1017/beq.2019.33

Vaara, E., Sonenshein, S., & Boje, D. (2016). Narratives as sources of stability and change in organizations: approaches and directions for future research. *Academy of Management Annals, 10*(1), 495–560. https://doi.org/10.5465/19416520.2016.1120963

Valentinov, V. (2014). The complexity–sustainability trade-off in Niklas Luhmann's social systems theory. *Systems Research and Behavioral Science, 31*(1), 14–22.

Valentinov, V., Verschraegen, G., & Van Assche, K. (2019). The limits of transparency: A systems theory view. *Systems Research and Behavioral Science, 36*(3), 289–300. https://doi.org/10.1002/sres.2591

Van Assche, K., Beunen, R., & Duineveld, M. (2013). *Evolutionary Governance Theory: An Introduction*. Springer.

Van Assche, K., Beunen, R., & Gruezmacher, M. (2024). *Strategy for Sustainability Transitions: Governance, Community and Environment*. Edward Elgar Publishing.

Van Assche, K., Beunen, R., Gruezmacher, M., Duineveld, M., Deacon, L., Summers, R., Hallstrom, L., & Jones, K. (2019). Research methods as bridging devices: Path and context mapping in governance. *Journal of Organizational Change Management, ahead-of-print* (ahead-of-print). https://doi.org/10.1108/JOCM-06-2019-0185

Van Assche, K., Greenwood, R., & Gruezmacher, M. (2022). The local paradox in grand policy schemes. Lessons from Newfoundland and Labrador. *Scandinavian Journal of Management, 38*(3), 101212. https://doi.org/10.1016/j.scaman.2022.101212

Van Assche, K., Gruezmacher, M., Summers, B., Culling, J., Gajjar, S., Granzow, M., Lowerre, A., Deacon, L., Candlish, J., & Jamwal, A. (2022). Land use policy and community strategy. Factors enabling and hampering integrated local strategy in Alberta, Canada. *Land Use Policy, 118*, 106101. https://doi.org/10.1016/j.landusepol.2022.106101

Van Assche, K., Verschraegen, G., & Gruezmacher, M. (2021). Strategy for collectives and common goods: Coordinating strategy, long-term perspectives and policy domains in governance. *Futures, 128*, 102716–102716. https://doi.org/10.1016/j.futures.2021.102716

Vanheule, S. (2016). Capitalist discourse, subjectivity and Lacanian psychoanalysis. *Frontiers in Psychology, 7*, 1948. https://doi.org/10.3389/fpsyg.2016.01948

Vanier, A. (2010). Fear, paranoia, and politics. *The Psychoanalytic Review, 97*(2), 215–229. https://doi.org/10.1521/prev.2010.97.2.215

Wahl-Jorgensen, K. (2019). *Emotions, Media and Politics*. John Wiley & Sons.

Yanow, D. (2000). *Conducting Interpretive Policy Analysis* (Vol. 47). Sage.

Zaretsky, E. (2008). Psychoanalysis and the spirit of capitalism. *Constellations, 15*(3), 366–381. https://doi.org/10.1111/j.1467-8675.2008.00497.x

Ziai, A. (2016). *Development Discourse and Global History: From Colonialism to the Sustainable Development Goals*. Taylor & Francis. https://doi.org/10.4324/978131 5753782

Žižek, S. (2002). *Did Somebody Say Totalitarianism?: Five Interventions in the (mis) use of a Notion*. Verso.

Žižek, S. (2006). *Interrogating the Real*. Bloomsbury Publishing.

Žižek, S. (2008). *Violence*. Picador.

Žižek, S. (2009). *In Defense of Lost Causes*. Verso. http://public.ebookcentral.proqu est.com/choice/publicfullrecord.aspx?p=5176960

Žižek, S. (2011). *Living in the End Times*. Verso Books.

10 Futures imagined and feared

Stories about the good community and how it should be governed, have a temporal dimension and a spatial dimension. The community to be governed lives somewhere, and while some networked and dispersed groups of course developed their own forms of governance, most existing configurations serve an area, and where governance is formalized, there will be administrative and political boundaries, which might or might not entirely overlap with the spatial boundaries of the community as a group. People belong to different groups, have layered identities, and administrative boundaries might reflect the main identification, or not.

Temporal dimensions are more complex (see also Chapter 2) as stories about past, present, and future mutually constitute each other, while commonly finding reference in space as well. Our discussions of legacy, memory, and history pointed out that the past is present in governance in many ways, and that the understanding of the present, and the visions of the future, the futures hoped for, feared, and deemed realistic, are more than tinged by the versions and traces of the past in governance and community, and the pre-occupations of the present, while the tools available to organize better futures are inherited from the past.

Governance can crystallize futures for communities, can trigger guided self-transformations, and can produce strategy for collectives and common goods (Kornberger, 2022). Strategy outside governance would be hampered by a lack of legitimacy from the start, and a lack of tools for coordination. The construction of futures in governance systems, if it takes place at all, is riddled with problems (MacKenzie, 2021). The process cannot be further from the image of rational decision-making projected by modernist governance systems. Hopes and fears, desire, a tangle of conflicted self-images, and complicated relations to miracle solutions offered by experts, to stories of success and failure with neighbors, do not just tinge but structure the process where some narratives gain credibility and others are easily dismissed (Gunder, 2016).

Many communities refrain from developing long-term perspectives and strategies trying to move in the direction of such imagined futures, and the

DOI: 10.4324/9781032696676-10

reasons for not trying are manifold and usually not related to resources or facts. Ideological meta-narratives on the good life, the "real" community, the "real" economy, the best form of democracy, and the appropriate forms and functions of governance are easily recognizable culprits. Productive fictions of unity, stability, harmony, and identity itself can hamper the development of futures and strategies that offer better adaptation to changing environments, and better expression and development of existing potential – to move in a direction compatible with real assets, values, and desires. Clinging to a present, to an imagined past, can easily obstruct the capacity to imagine better futures, or render those futures into repetitions of the past, into shallow adaptations, or into alternatives that err on the side of unproductive fantasy (Glynos, 2011; Van Assche et al., 2023). They can also exclude voices and perspectives that could open new avenues of self-reflection and reinterpretations of possible futures (Petriglieri & Petriglieri, 2020).

Enhanced reflexivity can shed a light on the origin of discourse and affect, latent and overt, structuring visions and strategies for the future, or highlighting their absence. The question is not only what the deeper drivers of imagined futures and their paths of implementation are, or what comes from where, why, and how. The issue is also which forms of sublimation might be at play, which transformations of motifs and motivations (Valdre, 2019), which productive fictions are involved, and to what extent they can be assessed in their effects on the legitimacy, viability, desirability, and inclusivity of the imagined future –and we come back to this in our final chapter.

Fantasy and reality testing

Fantasy is key in the construction and implementation of futures. Processes of collective transformation offer a vast playing field for fantasy, for the reflection of existing fantasy and the creation of new ones (Žižek, 1992, 2020), as desires can play out in the construction of desirable futures, in visions for the future, and derived strategies are supposed to be fantasy. Not every desire is incompatible with others, not every desire is unrealistic, and, as we know, our individual and collective versions of reality are always fantasy constructs. Not only is desire integrated into our sense of reality, but fantasy also gives us orientation by infusing perspective into each interaction with the world, and by glossing over the gaps within and between the symbolic and imaginary orders (Catlaw & Marshall, 2018; Eberle, 2019).

Fantasies about the future can be permeated by fantasies and affects relating to very different things. The imagined future can represent a clear break with something thoroughly despised in the current community; it can offer itself as an escape, a way to get rid, in the imagination, of enemies, of those who stand in the way of the imagined perfect community (McAfee, 2019; McGowan, 2017). The imagined future can take its shape from fantasy scenarios known from literature, from popular culture, while affects produced by those cultural products can be transferred to the community future (Hunt, 1989). Stories about the future can absorb features of fantasy scenarios about the future, but can absorb just as much stories about past or present, about other places, imagined places. The connection with the present and the past, with other domains of culture and society is thus multicolored, and these connections might be traceable easily or with great difficulty. Feelings that became amenable to articulation through poetry might have been translated into feelings about topics very different from what the poetry considered, for example, into feelings that led to policies and spaces detested by a silent majority. That majority might have formed through a mutual recognition of resentment and might creatively construct new futures from a narrow negative basis, from a very clearly shared dislike for a few policy positions, or an affective alienation in an environment associated with the powers that exist.

If in a process of visioning about the future, ample opportunities are given for participation, for free and creative thinking but also for the airing of grievances, more of what is happening in the collective conscious and unconscious can be gleaned by sensitive organizers, by leadership (see also next chapter). We would add that very careful consideration is needed, if such insight is welcomed, of the directions given to participants and to the overall framing of the exercise (Helling, 1998; Shipley & Michela, 2006). Is this about solving current problems? Is it a matter of forecasting and working from there? Or, is this about imagining and discussing possible futures, of better states of the community, in its physical, social, economic aspects? Different preoccupations, Master Signifiers, hopes and fears, desires and drives might become visible, and longer lines of questioning (why this, why that) might reveal more associations with repressed signifiers and disavowed emotions (Grigg, 1991; Wolf & Schwartz, 1962).

Processes of participatory visioning might have a therapeutic aspect, yet they are not a full analysis. In the next chapter we will discuss versions of self-analysis that can go further. As sessions are collective, that collective has to be trusted, and the process has to be trusted, including by implication, the organizers (de Vries et al., 2014; Shipley & Newkirk, 1998). If people do not trust government (or its consultants), if they do not see the value in discussing the long term, or the value of planning, or if they believe other residents present have reasons not to be open, or reasons to weaponize their own openness, the process will lead nowhere. It will reveal very little about what individuals or collectives are feeling and thinking. When the interpretation of self and environment is extremely dire, even thinking about the future, about alternatives is felt as irrelevant (Bennett, 1997; Van Assche et al., 2023).

Even if it proves possible to organize visioning in a manner that reveals more about the conscious and unconscious desires of the community, there still must be a phase of waking up, after fantasies are brought into the open. The waking up does not have to be a rude awakening, yet other interpretations of realities do need to enter the discussion, versions entertained by the self and by others present, and in a discussion that can function as reality testing (Driver, 2009; Dunlop & Radaelli, 2018). Such discussion can produce new, but more grounded and more widely shared fantasies about the future. When this happens, one can truly speak of visioning, as in the creation of a shared vision (Gruezmacher & Van Assche, 2022).

A visioning process has to be open-minded to function, and, as in psychoanalysis, moral judgment has to be suspended for the moment (Hillier, 2003). This, of course, can be very difficult in the public sphere, as any public appearance is by definition coded, inhibited, is a performance. Not everything can be said, can be admitted, as cultural and more formal rules do not allow this, and as organizers might be legally obliged to respond when certain forms of speech appear. Nevertheless, what is disavowed or repressed, what is not allowed to be expressed, or what one does not want to observe and recognize in others, in the community, has the tendency to rear its head somewhere else (Petriglieri & Petriglieri, 2020; Phillips, 2024). Anger can find new causes and enemies, desire can find new expressions, and utopian inclinations might be transferred to the neighborhood level.

The game of formal politics, even at the most local level, is not sufficient to reveal and guide collective desires as pertaining to community futures (Cederström, & Hoedemaekers, 2010). This is a matter of simple observation, of familiarity with the history of politics, planning, and administration (Gunder, 2005, 2010; Gunder & Hillier, 2004). Systems adjust themselves to changing desires and requirements, about the good community and the role of governance there, yet, in whatever version, formal politics is not a sufficient arena and set of mechanisms to translate aspirations for the future into actionable strategies (Hillier, 2000; Žižek, 1992). Hence our interest in governance, as governance is always there, always broader than politics, and in opening up collective thinking and decision-making for the potentiality of many channels (Pierre & Peters, 2020; Savski, 2020). Some versions of governance will not be interested in visioning, as some communities might not be interested, but the collective interest can be awakened if people are made aware that governance is capable of doing more. If a version of visioning is accepted, if there is a shared understanding that it is possible to identify public goods, collective desires, and craft a shared future, the place to do such thing can still be contested or unclear (Czarniawska-Joerges, 1999; Jarzabkowski, 2005). What is realistic in a community, has to transpire from a thorough self-understanding in the community and governance system, and this entails, again, a deep familiarity with the governance path and the myriad links between governance and community.

Moments, spaces, organizational forms, types of guidance, and the structuring of a visioning process that might be workable can only be identified through such familiarity. Consultants or other outsiders offering ready-made formulas can be helpful, by speeding up learning, and stopping excessive self-doubt, yet represent a risk of not grasping the intricacies of governance and community, of not contextualizing the process (Alvesson et al., 2009). Some would counter that the process itself can bring familiarity, can produce contextualization, yet the time devoted usually does not allow for much learning (by consultants), while the openness of participants and the effectiveness of the process are most likely determined by factors not observed by consultants (or visiting academics) and not built into the procedure. In short, only insiders are likely to succeed in assessing when and how visioning might work, with or without outsiders involved (Helling, 1998). At the same time, the visioning itself, in its therapeutic aspects, can be conducted by outsiders, as long as they are trusted, and here, a perception of neutrality, of having no stake and no position in the community, might make it easier to function as a Freudian blank slate, or as a screen for the projection of ideals, hopes, and fears (Muhr & Kirkegaard, 2013).

Even in such a psychoanalytically inspired version of visioning, limits will have to be observed, and, as government is often involved in organizing them, the symbolic order is always present. Speaking freely when closely watched by the Big Other, and by many little others, with whom symbolic competition threatens to trigger spiraling emotions, and a conception of policymaking for the future as a zero-sum game is certainly challenging (Gunder, 2003). Then again, spaces can be created where more freedom is explicitly allowed, that is, situations where distance is maintained from the more intimidating aspects of the symbolic order. Free speech, however, does need to be accompanied by an injunction for self-reflection, which can be combined with reflection by others, after space has been given for the articulation of fantasy about the future. The symbolic order, in this process, can be both constraining and enabling, as it can open up possibilities for a consideration of possible futures disconnected from imaginary identifications and competition, where others in the community, and in the room, can disappear into a nonthreatening background until the moment they have to be considered again in a discussion of what constitutes public goods and desirable futures (Frosh, 2001a; Žižek, 2002).

Distinguishing realistic futures, structured or tinged by fantasies that might be compatible with each other and with a stable interweaving of the Symbolic, Imaginary, and Real from fantasies which will unravel in the face of present and future realities is not an easy matter for the observer of governance and for the participants in governance. Still, such distinctions can be made and must be made for governance to continue its coordinating function, for the community to maintain its cohesion, and for the hope in better futures to stay alive (Arlow, 1969; Kornberger & Clegg, 2011).

Merely opening discussions on alternative futures, on strategizing for the future, or just on the future can already trigger fears, irritation, unease, and alienation (in the sense of the Freudian *unheimische*). Unrealistic or unacceptable futures can be articulated just to annoy others, to mock or disrupt the system, to cement the unity of the rebellious group. And that group can represent a majority, or create a majority, through its opposition to forms of meddling, to governance processes felt as an imposition, or as a threat to local identity (H. T. Miller, 2012; Pohl & Swyngedouw, 2023). On the positive side, all such feelings can be productive as well. Aggression can sharpen critical thinking and can serve to separate from old identities, as well as underpin attachment to them (Hagman, 1995). Love for the existing order of things can sublimate into love of the things themselves, or love of abstracted features of the current community, which could be more fully actualized in future versions of the place. Alienation, as in the denaturalization of perspective, can set people on the path to new encounters with other perspectives and reevaluation the old ones.

Reality testing, therefore, cannot be omitted, but can also not be seen as a clear step, following a simple procedure. Creating distance from current realities, establishing the interweaving of desire and cognition, and recognizing the dominance of certain Master Signifiers can all serve to deconstruct what counts now as reality, as good governance to discover blind spots and less than productive fictions. Reality testing, therefore, cannot be a check on conformance with an objective reality, but rather a check of the *effects* of current reality constructions, and the effects of realities consciously constructed later and better. This means that it is a never-ending process in governance and community, intensified in moments of visioning and in moments of collective choice.

Hopes and dreams

In the construction of futures, hope can be invested, and the lack of detail available to us about future events and their connectivity makes it easier to project into the future what might not be recognized as realistic right now (Gunder, 2016). Projection into the future can serve several functions here. Reinforcing the current self-image can be a factor, but so can an overinvestment or lack of investment in the materialization of a particular narrative (H. T. Miller, 2012). Overinvestment in what is important now can make communities assume it will always be important and can instigate the making of long-term plans revolving around the current concern or asset. Lack of investment refers to the use of the long term to not take decisions, to postpone something forever. None of these mechanisms is helpful for communities to explore which desirable and realistic futures are available to them. Using the future as a rhetorical trash can, or as an extension of the present does not offer much insight into current or future realities, yet begs the question as to the reasons for this lack of interest in either present or future.

Moreover, what appears as future reality may appear as a dream for others, while realities can acquire a dream-like quality when confronted with unexpected events, or when a shift in affective investment in different narratives occurs (Orr, 2023; Žižek, 1992). Such shifts can take place through metaphoric or metonymic association. An explosion of a pipeline in the far north, an attack on a playground in Russia, an earthquake in Mongolia, a movie about all those things can erode the feeling that hopes might become reality, by undermining feelings of security and system stability, independent of any rational risk calculation.

More poignantly perhaps: a visit to a place that is possibly unconsciously recognized as similar, can lead to an accidental encounter, which might not be representative of life, of quality of life in the place, but nevertheless thoroughly reframes one's own feeling about the future (Thrift, 2004). It might be that an *objet petit a* loses its luster, that something unconsciously central to the quality of the future is revealed as irrelevant (Bistoen, 2016). Desire, and hope, can move somewhere else. Indeed, the opposite move is just as likely, with previously invisible places, people, and situations seemingly irrelevant for an imagined future, suddenly becoming pregnant with meaning and affect. A drab neighborhood can reveal an unexpected quality of life, can reveal the futility of an obsession with counting trees as indices of sustainability or design quality. While this might feel like a discovery, and while new arguments, previously overlooked, might indeed come into the picture, what really caused the shift in perspective, in understanding and affect, can be quality, an intensity, an event, a temporality, a detail that can only be discovered in hindsight, by reconstructing associative chains (Freud, 2003, 2015, 1983; Gentile, 2020). These might exist in the unconscious of individuals visiting the place, who then influence the home front, and they might spread through cultural products, through shared experiences (Frosh, 2001b)

Communities can dream big, and they can have relinquished their dreams a long time ago. They might also believe that only individual dreams are worth dreaming and pursuing (McGowan, 2017). Ideals and high hopes can mobilize a community, can galvanize it, yet when confronted with disappointment or with the specter of conflict or the daunting task of managing complexity, ideals and hopes can also immobilize it (Barratt, 1985). It is entirely possible to cling to hopes and ideals without ever actively pursuing them, which is a truth for individuals and for communities. The fear of failure can be immobilizing, as is the fear of revealing oneself to be inadequate or just different from what one prefers to think.

Thus, dreams and hopes can unite the community, and they can divide it (Renström et al., 2023). They can bring a community into being through sharing an optimistic vision of the future, and they can divide when those visions are hard to realize, or when fractures appear where new experiences lead to diverging interpretations. High hopes can divide the community when different affective investments come to light, and possibly different economic positionings – winners and losers might become visible, whereas initially only winners were declared (Glynos & Stavrakakis, 2010; Scott, 2020). Hopes and dreams might conceal Master Signifiers that were never shared in the community, and different instances of sublimation, rooted in contrasting desires (Kim et al., 2013; Moyaert, 2007).

Hopes and dreams can stop critical thinking if they are taken as realistic, before there is any assessment of what realism could mean. If hopes and dreams are connected to an understanding of self and environment that is fixed by ideology or other rigid sets of Master Signifiers, or to unexamined and unobserved desires, this increases the chance that fantasy scenarios take over, which give free play to desires and truths that were never questioned, and never given the chance to be confronted by alternatives (McAfee, 2022; McGowan, 2021). Yet, they can do the opposite, as when they are grounded in a firm self-understanding, a shared assessment of assets that might be of value for future environments. In this case, hope can guide observation and self-observation: if this is where we want to go, and we believe this is possible, then we must look for signs of one thing or another in the world around us, and make sure that we ourselves are equipped to do whatever it is that is needed to move forward (Van Assche et al., 2024).

Hopes and dreams can thus be productive, structuring productive fictions and providing the motivations to make them real, while the opposite always remains possible. Governance systems, when trying to help communities construct futures and bring them closer, can inspire hope, and share dreams of the future. It might be that elsewhere communities already share aspirations and excitement about a new version of the community that is already possible. Yet in other cases it can truly be described as co-production, meaning that not only the imagined future but also the hopes and other positive affects invested in it are the product of a process of collective visioning, of a process structured by governance where new hope is elicited (Karadima, 2022). Hope is for us, therefore, both a belief and an affect.

Leadership can place itself in the role of reflecting hopes and dreams, uncritically accepting them and magnifying them, or it can take different positions (Alvesson, 2011; Alvesson & Spicer, 2012). Inspiring visions can come from leadership, and communities can identify with them. Hopes can be projected onto the leadership itself or onto the futures produced, and the projection can shift, sometimes imperceptibly between those. People can forget that leadership was responsible for a vision now integrated into the sense of self (Abse & Jessner, 1962; Maccoby, 2004). Leadership can thus even be attacked in the name of what it produced and what is now highly appreciated, by splitting the image of leadership from anything positive that is now part of the self-identification.

We are all familiar with movements in the other direction, where leadership, that we can call charismatic (following Max Weber; see Iordachi, 2004) can get away with any ideological shift, any incoherent vision, as the investment in the person is a guarantee for the future. Successes one is not really responsible for can be claimed more easily, as a critical reception of leadership rhetoric is absent. In charismatic leadership, rather than splitting, one can speak of expansion of the leadership image, sometimes to the point where the leader is the only stable point of reference.

Fear and anxiety

As the future is unknown, and as this entails uncertainty about the value of present investment and the meaning of present activities, relating to the future is inherently linked to the management of anxiety (Kinnvall & Svensson,

2022). One way to deal with this is to use the collective as a shelter, which can reassure us that nothing will ever change (Žižek, 2002, 2006). A different approach is to convince ourselves as individuals that we can control our own destiny, and that the community just has to create a few basic conditions (Penz & Sauer, 2019). A third main avenue is to construct communities that offer a promise, through governance, to improve conditions and solve problems (a premise stretching beyond modernism and beyond democracy) (Latour, 2004). Finally, we can mention religious (or religious-like) communities that offer certainty about the afterlife and that render the vagaries of the future less relevant (McCormick, 2011).

Governance paths tend to fit one of these main avenues, and internal diversity in larger communities allows for the coexistence of different varieties. Anxiety, however, cannot be eradicated (see Chapter 4), as nothing ever turns out exactly as intended, and as the strategies to explain away the difference, including the performance of success in governance, and ritualized action and discourse, have their limits (Pressman & Wildavsky, 1984; Van Assche et al., 2012). Anxiety, moreover, appears because many other issues translate into anxiety. Doubt, repressed ideas and emotions, internal contradictions, and prolonged silences, either on particular topics or as a general absence of public discourse, can trigger anxiety, while anxiety, once observed, spreads like wildfire (Toff & Nielsen, 2022). Incomplete understanding of issues, difficulty in coming to shared identity narratives or to broadly supported policies, and observation of conflict, even far away, can all translate into anxiety. A sense of self that is unstable, an uncertainty about connection with others, with the world, can generate intense anxiety (Zevnik, 2017).

Anxiety can trigger fear, and vaguely defined fear can disperse into anxiety around many more topics and areas of life than what was initially, if only vaguely, defined. Freud pointed out how both fear and anxiety travel unconsciously, and can attach to new objects and situations (Freud, 2013, 2014, 2020). Anxiety can be managed by turning it into fear; fear of something that is not supposed to be feared can metamorphose into a conscious fear of something else. Fear of open spaces, if we follow Freud, can be a displacement of fears or desires in ourselves that we do not want to confront. Fear of wasting time or money can stem from a fear of not being adequate in the eyes of a person of authority, which can be a manifestation of the Big Other, and possibly fear of not understanding, or not believing enough the narrative that tells us how time and money should be spent and what constitutes a waste (Hook, 2008). From there, the narrative can come to stand for a wasted life, or, for a community that does not live up to its promise, or focuses on the wrong things (Pusca, 2007).

This last example illustrates how ideologies, in this case neoliberalism, come with their own anxieties and their own ways of crystallizing around (conceptual) objects and translating them into fear. Fear can be less destabilizing than anxiety, as it can be given a referent, a more stable meaning and position in life. If fear turns into anger, which is a common occurrence, since few communities want to admit having fear because it does not fit dominant discourses, and since governance does not want to question those discourses, a scapegoat responsible for the fear can be delineated (Savun & Gineste, 2019). Getting rid of the scapegoat then offers hope of allaying fear, and, in a more official explanation, solving a problem. That problem then has to be connected, in surface discourse, to the scapegoat, so its removal can be more convincing – can be proposed as a solution (Vanier, 2010). In other words, in the worlds of policy and governance, it is often not only a detour of affect but also a construction of problems and objects that is implied in scapegoating.

Hate and anger can be cultivated, and they can be remembered and revived. Machiavelli spoke of the long memory that makes "old injustices" politically alive again (Mansfield, 1998). Conflicts can start with anger, but also for many other reasons, yet the conflict itself can leave a long trail of anger and fear. It can solidify internally and externally exclusive identities, and polarizations that can become parasitic on other relations, other domains of policy and aspects of life. Conflicts can, in other words, create and cement categories and distinctions that survive their utility for a long time (Luhmann, 1995). Identities that were more layered and contextual, and altogether less relevant before a conflict, can suddenly become essentializing and polarizing (Magris, 2011). An ethnic identity can be a background factor, until fears are raised, until a conflict forces members to make it important or abandon it altogether (Frosh, 2001b; Hoggett, 2015). Associating with that ethnicity transfers doubts about the purity of those people crossing boundaries now deemed defining for the community. Crossing the boundary aggravates the threat to the community (Laclau, 2005).

Since the early twentieth century, modernist theories of governance have suggested that the future does not pose too many problems. The epistemology of modernism, the belief in the rationality of humans and the unity of truth, as well as the stability of identities and the predictability of needs conspired to create the impression that optimizing the territory, and solving societies' problems was a technical exercise (Allmendinger, 2017; Hillier, 2003). Despite decades of theoretical innovation, and despite the insights of Friedrich Nietzsche, Karl Marx, and Freud, which predated the zenith of modernism in the 1950s and 1960s, modernism is not dead. It refuses to die, and we have discussed some of the reasons in earlier chapters. What deserves repeating here, is the desire to manage anxiety about the future, and, simultaneously, anxiety about the object of the community itself. If we can fix identities, can objectively define needs, then we do not have to consider desires, and the instability of discourse and desire (Cossman, 2013; Pivnick & Hassinger, 2022; Pottage & Mundy, 2004).

A modernist perspective on governance and the world cannot banish uncertainty, but this can be quantified and tamed under the name of risk. Risk is the chance that something bad will happen, and for a modernist, especially one with a capitalist bent, it is possible to list the bad things that could happen and calculate the risk they pose (Fischer, 2000; Luhmann, 2017). This is expected to banish anxiety, yet the relentless search for new risk objects and improved risk calculations generates not only a continuous anxiety about risks being missed and miscalculated, but also a cycle of fear that can attach to ever-new risk objects (Scherz, 2022). An institutionalized risk gaze is bound to find new risk objects, which can then attract fear, that can be used as an argument to validate the prominence of the risk gaze (Gunn & Hillier, 2014). The risk perspective on governance can produce new blind spots for the positives, a lack of interest in discussing what is of value for the community, and in life (Schehr, 2005). And it can, unfortunately, reduce our desire to change the world to create new public goods, and our curiosity to learn and rethink the quality of life.

In a neoliberal ideology, when things happen to individuals, reconfiguring governance and community to make different things happen is less central to the picture. What remains is a world of individual opportunities and individual risks. Constructing a collective future is not believed to be possible or desirable, and imaginary futures are reduced to individual fantasies, or to mercenary forecasts, revolving around calculations of risk and benefit (Hacking, 1990; Rose, 1991). This example illustrates how a concept like risk can become central in governance because of a shared master narrative or ideology, which can permeate society, and which can work as a factor delimiting the possibilities to imagine and create better collective futures. Ideology can serve to imagine collective desires and collective goods as non-existent, or only to become visible through discrete decisions on services or on a limited repertoire of topics accepted as belonging to the public domain – matters of war and peace, freedom of expression, of ease of doing business (Mannheim, 2013; McGowan, 2017).

From a psychoanalytic point of view, the modes of self-limitation that are at work in a community, limiting its understanding of itself, its environment, and the futures available to it, are of the utmost importance. From a political point of view, they are no less important. From the previous passages and chapters, it might be clear that fear and anxiety, but also hopes and dreams can be productive as well as self-limiting (Lacan, 2014; McSwite, 1997). Indifference, anger, obsession, guilt and shame, the conscious and unconscious can all be present, and can limit self-knowledge and a free exploration of possible and desirable futures (Czarniawska-Joerges, 2008). They can be mobilized by those who are not interested in such an exploration, those who are keen to keep things as they are, or to promote a course benefiting themselves.

Narratives eliciting fears, conveying doubt, and attributing guilt or disavowing it can be just as detrimental as narratives promoting the limiting of ideas on good governance, the good community, and public goods (Glynos, 2010; Žižek, 2019). The issue is not one of choice between big or small government, but between one where choices remain possible for the construction of collective futures, and one where the whole endeavor is a priori deemed impossible, even immoral – since it infringes on freedoms and wastes public money (Lambert, 2019).

Concluding: strategy as a possibility

Among the more pernicious narratives limiting the self-creation of communities are those insisting that even long-term perspectives have no place in governance. Among the more deplorable situations, one can count one where, for whatever reason, long-term perspectives vanished or were never present (Underdal, 2010). We are not speaking theory here, as many communities never developed the capacity to entertain ideas about the longer-term future and derive collectively binding decisions from them. In some cases, stories about the future appear and disappear, leading or not leading to a decision, yet not altering governance in such a manner that considering the long term becomes a regular practice, or that decisions for the long term remain binding over a longer period of time (Luhmann, 1987, 1989). Futures can be disavowed, in particular and in general and, as we discussed earlier this chapter, this can stem from structural choices in the governance system but also from contingent events and ad hoc adaptations.

We argued for participatory forms of visioning, which, though not always possible, can make a community reflect on its own narratives and affects about the future, before coming to a shared vision. We can safely say that such vision can only hope to be a productive fiction, as the sharing cannot be ascertained, and grasping what it is exactly that is desirable for all is a challenge. We can also say that a process of visioning that rests on a patient reflexivity, on a careful unearthing of expectations, of assumptions and affect, can increase the chances that the fiction will be a productive one, by embedding it in the narrative and affective landscape of the community. The process can help connect visions to existing stories, hopes, and fears, but in such a way that fantasy can be grounded, that problematic narratives can be exposed and affective economies better understood. Creating space, in governance and in public discourse for considerations about the future, for the creation of imagined futures, thus must be combined with the cultivation of reflexivity (Van Assche et al., 2021).

Sometimes, this might take the character of therapy, and in the following chapter we ask ourselves what this could look like. The previous paragraphs already give an indication, yet do not dwell on what serious cases might be, and whether other forms of analysis might be necessary. Here, we limit ourselves to a plea for entertaining the possibility of collective strategy. Such strategy does not have to take the form of a comprehensive policy, nor is it obliged to rest on comprehensive studies (Latour, 2004). It would have to follow from a self-analysis and build on the articulation of a vision. What makes it a strategy is the reformulation of that vision as a narrative, into a language that is recognizable for governance. A strategy, in our view, is possible for most governance systems; it is both a narrative and an institution, both a (reformulated) vision and a tool to coordinate action and decision-making (Van Assche et al., 2020). This coordinative dimension will entail the coordination or integration of other institutions, other policies, plans, and laws, either existing ones or new (Czarniawska, 2014; Kornberger, 2022). The narrative side will ensure a linkage to power/knowledge configurations in governance, and to the aspirations in the community at large, while the coordination of other policies is necessary to fit into and mobilize actor/institution configurations. The combination of the two dimensions increases the chances of implementation, by intervening in both thinking and organizing. Strategy embodies the unity of the distinction.

Not all communities are ready to strategize, and self-analysis can help assess the readiness. Not every community is even ready to envision the longer-term future in a way that takes into account all options available to it. Communities can be stuck in a past, in a present, in a future, in unique ways that can only get unstuck after the community acquires a deeper insight in itself. The self-knowledge can extend to its own governance path, its desires and drives, its power/knowledge configurations, but also unacknowledged affects, Master Signifiers, and imaginary identifications. Not all communities will be interested in these concepts, or in any self-analysis beyond what is perceived to be the objective needs of the day (Groarke, 2013; Gunder, 2016). Thus, as with individuals, communities most in need of (self-) analysis might be the least inclined to see that need. If we take democratic principles literally, communities, if they so choose, can be as blind as they want, as stuck as they want, in a phase of development that does not do justice to their potential (Glad, 2002; Weinberg, 2006). They have the right to remain lodged in an impasse where uncomfortable truths might be felt but without surfacing, where a potential future might be known but disavowed, bereft of any affective investment and thus felt reality.

Where communities do embark on strategy, without a preceding self-analysis, it might lead to a repetition of moves, to overly critical treatment of or blind belief in proposals coming from external advisers and great difficulties in reality testing. Strategizing can then easily become another exercise in the production of paper tigers, and negative feelings and ideas about governance, about working on a collective future can be reinforced (Panizza & Stavrakakis, 2020). Moreover, affects can surface that cannot be handled in the moment, and tath can threaten the cohesion of the strategy process, its felt legitimacy, and, in the worst case, the cohesion of the community trying to rethink itself (Mintzberg, 1989).

We add that the timelessness of the traumatic and the repressed can alter the temporality and reality of future visions; some things do not change, yet cause unexpected symptoms later (Freud, 2020). This in turn might represent a separate set of boundaries for community strategy. Thus, coming to grips with trauma before embarking on anxiety-provoking visioning and strategy exercises, however difficult it may be, seems far more preferable to unleashing it in a process and on a community not prepared to handle it (Salberg & Grand, 2024; Schultz et al., 2016).

What could happen when a community, with a difficult history, is willing to examine itself, to bring out what is left unsaid and what has remained unobserved, and from there look again at the future, we explore in our next and last chapter.

References

Abse, W., & Jessner, L. (1962). The Psychodynamic Aspects of Leadership. In S. Graubard, & G. Holton, (Eds.). *Excellence and Leadership in a Democracy* (pp. 76–93). Columbia University Press.

Allmendinger, P. (2017). *Planning Theory* (3rd ed., p. 346). Macmillan Education UK. https://books.google.ca/books?id=XQAoDwAAQBAJ

Alvesson, M. (2011). Leadership and Organizational Culture. In D. Schedlitzki, et al. (Eds.), *The SAGE Handbook of Leadership* (pp. 151–164). Sage.

Alvesson, M., Bridgman, T., & Willmott, H. (2009). *The Oxford Handbook of Critical Management Studies*. Oxford Handbooks.

Alvesson, M., & Spicer, A. (2012). A stupidity-based theory of organizations. *Journal of Management Studies, 49*(7), 1194–1220. https://doi.org/10.1111/j.1467-6486.2012.01072.x

Arlow, J. A. (1969). Fantasy, memory, and reality testing. *The Psychoanalytic Quarterly, 38*, 28–51. www.tandfonline.com/doi/abs/10.1080/21674086.1969.11926480

Barratt, B. B. (1985). Psychoanalysis as critique of ideology. *Psychoanalytic Inquiry, 5*, 437–470. https://doi.org/10.1080/07351698509533598

Bennett, O. (1997). Cultural policy, cultural pessimism and postmodernity 1. *International Journal of Cultural Policy, 4*, 67–84. https://doi.org/10.1080/10286639709358063

Bistoen, G. (2016). The Lacanian Concept of the Real in Relation to Politics and Collective Trauma. In G. Bistoen (Ed.), *Trauma, Ethics and the Political beyond PTSD: The Dislocations of the Real* (pp. 104–130). Palgrave Macmillan UK. https://doi.org/10.1057/9781137500854_6

Catlaw, T. J., & Marshall, G. S. (2018). Enjoy your work! The fantasy of the neoliberal workplace and its consequences for the entrepreneurial subject. *Administrative Theory & Praxis, 40*(2), 99–118. https://doi.org/10.1080/10841806.2018.1454241

Cederström, C., & Hoedemaekers, C. (2010). *Lacan and organization*. MayFlyBooks.

Cossman, B. (2013). Anxiety governance. *Law & Social Inquiry, 38*(4), 892–919. https://doi.org/10.1111/lsi.12027

Czarniawska, B. (2014). *A Theory of Organizing*. Edward Elgar Publishing.

Czarniawska-Joerges, B. (1999). *Writing Management: Organization Theory as a Literary Genre*. Oxford University Press.

Czarniawska-Joerges, B. (2008). *A Theory of Organizing*. Edward Elgar.

de Vries, J., Roodbol-Mekkes, P., Beunen, R., Lokhorst, A. M., & Aarts, N. (2014). Faking and forcing trust: The performance of trust and distrust in public policy. *Land Use Policy, 38*, 282–289. https://doi.org/10.1016/j.landusepol.2013.11.022

Driver, M. (2009). Struggling with lack: A Lacanian perspective on organizational identity. *Organization Studies, 30*(1), 55–72. https://doi.org/10.1177/0170840608100516

Dunlop, C., & Radaelli, C. (2018). The lessons of policy learning: Types, triggers, hindrances and pathologies. *Policy & Politics, 46*(2), 255–272. https://doi.org/10.1332/030557318X15230059735521

Eberle, J. (2019). Narrative, desire, ontological security, transgression: Fantasy as a factor in international politics. *Journal of International Relations and Development*, *22*(1), 243–268. https://doi.org/10.1057/s41268-017-0104-2

Fischer, F. (2000). *Citizens, Experts, and the Environment: The Politics of Local Knowledge*. Duke University Press.

Freud, S. (2003). *The Uncanny* (H. Haughton, Ed.; D. McLintock, Trans.; Illustrated edition). Penguin Classics.

Freud, S. (2013). *The Problem of Anxiety*. Read Books Ltd.

Freud, S. (2014). *Wit And Its Relation To The Unconscious*. Routledge. https://doi.org/10.4324/9781315830759

Freud, S. (2015). *Beyond the Pleasure Principle* (J. Miller & M. C. Waldrep, Eds.; p. 64). Dover Publications.

Freud, S. (1983). Creative Writers and Daydreaming. In E. Kurzweill, & W. Phillips (Eds.), *Literature and psychoanalysis* (pp. 19–28). Columbia University Press. https://doi.org/10.7312/kurz91842-003

Freud, S. (2020). *The Interpretation of Dreams: The Psychology Classic*. John Wiley & Sons.

Frosh, S. (2001a). On reason, discourse, and fantasy. *American Imago*, *58*(3), 627–647. https://doi.org/10.1353/aim.2001.0013

Frosh, S. (2001b). Psychoanalysis, Identity and Citizenship. In N. Stevenson (Ed.), *Culture and Citizenship* (pp. 62–73). Sage.

Gentile, J. (2020). Time may change us: The strange temporalities, novel paradoxes, and democratic imaginaries of a pandemic. *Journal of the American Psychoanalytic Association*, *68*(4), 649–669. https://doi.org/10.1177/0003065120955120

Glad, B. (2002). Why tyrants go too far: Malignant narcissism and absolute power. *Political Psychology*, *23*(1), 1–2. https://doi.org/10.1111/0162-895X.00268

Glynos, J. (2010). *?Lacan at Work?* (C. Cederstrom & C. Hoedemaeker, Eds.; pp. 13–58). MayFly Books. https://repository.essex.ac.uk/4085/

Glynos, J. (2011). On the ideological and political significance of fantasy in the organization of work. *Psychoanalysis, Culture & Society*, *16*(4), 373–393. https://doi.org/10.1057/pcs.2010.34

Glynos, J., & Stavrakakis, Y. (2010). Politics and the unconscious – An interview with Ernesto Laclau. *Subjectivity*, *3*(3), 231–244. https://doi.org/10.1057/sub.2010.12

Grigg, R. (1991). Signifier, Object, and the Transference. In E. Ragland-Sullivan & M. Bracher (Eds.), *Lacan and the Subject of Language (RLE: Lacan)*. Routledge.

Groarke, S. (2013). *Managed Lives: Psychoanalysis, Inner Security and the Social Order: Psychoanalysis and the Administrative Task*. Routledge. https://doi.org/10.4324/9781315880150

Gruezmacher, M., & Van Assche, K. (2022). *Crafting Strategies for Sustainable Local Development*. InPlanning.

Gunder, M. (2003). Passionate planning for the others' desire: An agonistic response to the dark side of planning. *Progress in Planning*, *60*(3), 235–319. https://doi.org/10.1016/s0305-9006(02)00115-0

Gunder, M. (2005). Lacan, planning and urban policy formation. *Urban Policy and Research*, *23*(1), 87–107. https://doi.org/10.1080/0811114042000335287

Gunder, M. (2010). Making planning theory matter: A Lacanian encounter with Phronesis. *International Planning Studies*, *15*(1), 37–51. https://doi.org/10.1080/13563471003736936

Gunder, M. (2016). Planning's "Failure" to ensure efficient market delivery: A Lacanian deconstruction of this neoliberal scapegoating fantasy. *European Planning Studies*, *24*(1), 21–38. https://doi.org/10.1080/09654313.2015.1067291

Gunder, M., & Hillier, J. (2004). Conforming to the expectations of the profession: A Lacanian perspective on planning practice, norms and values. *Planning Theory & Practice*, *5*(2), 217–235. https://doi.org/10.1080/1464935041000169176

Gunn, S., & Hillier, J. (2014). When uncertainty is interpreted as risk: An analysis of tensions relating to spatial planning reform in England. *Planning Practice & Research*, *29*(1), 56–74. https://doi.org/10.1080/02697459.2013.848530

Hacking, I. (1990). *The Taming of Chance*. Cambridge University Press.

Hagman, G. (1995). Mourning: A review and reconsideration. *The International Journal of Psycho-Analysis*, *76*(5), 909.

Helling, A. (1998). Collaborative visioning: Proceed with caution!: Results from evaluating Atlanta's Vision 2020 project. *Journal of the American Planning Association*, *64*(3), 335–349.

Hillier, J. (2000). Going round the back? Complex networks and informal action in local planning processes. *Environment and Planning A*, *32*, 33–54.

Hillier, J. (2003). `Agon'izing over consensus: Why Habermasian ideals cannot be `Real'. *Planning Theory*, *2*(1), 37–59. https://doi.org/10.1177/147309520300 2001005

Hoggett, P. (2015). *Politics, Identity and Emotion*. Routledge. https://doi.org/10.4324/9781315632704

Hook, D. (2008). Absolute other: Lacan's 'Big other' as adjunct to critical social psychological analysis? *Social and Personality Psychology Compass*, *2*(1), 51–73. https://doi.org/10.1111/j.1751-9004.2007.00067.x

Hunt, J. D. (1989). *The Figure in the Landscape*. Johns Hopkins University Press. https://doi.org/10.56021/9780801817953

Iordachi, C. (2004). *Charisma, politics and violence: The legion of the "Archangel Michael" in inter-war Romania*. Trondheim Studies on East European Cultures and Societies, nr 15. NTNU, Trondheim.

Jarzabkowski, P. (2005). *Strategy as Practice: An Activity Based Approach*. Sage.

Karadima, D. (2022). Design and Commons: A Lacanian Approach. In G. Bruyns & S. Kousoulas (Eds.), *Design Commons: Practices, Processes and Crossovers* (pp. 223–238). Springer International Publishing. https://doi.org/10.1007/978-3-030-95057-6_12

Kim, E., Zeppenfeld, V., & Cohen, D. (2013). Sublimation, culture, and creativity. *Journal of Personality and Social Psychology*, *105*(4), 639–666. https://doi.org/10.1037/a0033487

Kinnvall, C., & Svensson, T. (2022). Exploring the populist 'mind': Anxiety, fantasy, and everyday populism. *British Journal of Politics and International Relations*, *24*(3), 526–542. https://doi.org/10.1177/13691481221075925

Kornberger, M. (2022). *Strategies for Distributed and Collective Action Connecting the Dots*. Oxford University Press USA - OSO. http://public.eblib.com/choice/PublicFul lRecord.aspx?p=6836917

Kornberger, M., & Clegg, S. (2011). Strategy as performative practice: The case of Sydney 2030. *Strategic Organization*, *9*(2), 136–162. https://doi.org/10.1177/14761 27011407758

Lacan, J. (2014). *Anxiety: The Seminar of Jacques Lacan, Book X* (J.-A. Miller, Ed.; A. R. Price, Trans.; 1st edition). Polity.

Laclau, E. (2005). *On Populist Reason*. Verso.

Lambert, A. (2019). Psychotic, acritical and precarious? A Lacanian exploration of the neoliberal consumer subject. *Marketing Theory, 19*(3), 329–346. https://doi.org/10.1177/1470593118796704

Latour, B. (2004). *Politics of Nature: How to Bring the Sciences into Democracy*. Harvard University Press. https://doi.org/10.4159/9780674039964

Luhmann, N. (1987). *Soziale Systeme: Grundriss einer allgemeinen Theorie*. Suhrkamp.

Luhmann, N. (1989). *Ecological Communication*. University of Chicago Press.

Luhmann, N. (1995). *Social Systems* (Vol. 1). Stanford University Press Stanford.

Luhmann, N. (2017). *Risk: A Sociological Theory*. Routledge. https://doi.org/10.4324/9781315128665

Maccoby, M. (2004). Why people follow the leader: The power of transference. *Harvard Business Review, 82*(9), 76–85, 136.

MacKenzie, M. K. (2021). There is no such thing as a short-term issue. *Futures, 125*, 102652–102652. https://doi.org/10.1016/j.futures.2020.102652

Magris, C. (2011). *Danube*. Random House.

Mannheim, K. (2013). *Ideology and Utopia*. Routledge. https://doi.org/10.4324/9781315002828

Mansfield, H. C. (1998). *Machiavelli's Virtue*. University of Chicago Press.

McAfee, N. (Ed.) (2019). Democratic Imaginaries. In Fear of Breakdown: Politics and Psychoanalysis (pp. 77–104). Columbia University Press. https://doi.org/10.7312/mcaf19268-008

McAfee, N. (2022). Public Philosophy and Deliberative Practices. In L. C. McIntyre, N. McHugh, & I. Olasov, (Eds.), *A Companion to Public Philosophy* (pp. 134–142). John Wiley & Sons, Ltd. https://doi.org/10.1002/9781119635253.ch14

McCormick, J. P. (2011). *Machiavellian Democracy*. Cambridge University Press.

McGowan, T. (2017). *Capitalism and Desire: The Psychic Cost of Free Markets*. Columbia University Press. https://doi.org/10.7312/mcgo17872

McGowan, T. (2021). Self-Destruction and the Natural World. In C. Burnham & P. Kingsbury (Eds.), *Lacan and the Environment* (pp. 273–293). Springer International Publishing. https://doi.org/10.1007/978-3-030-67205-8_15

McSwite, O. C. (1997). Jacques Lacan and the theory of the human subject: How psychoanalysis can help public administration. *American Behavioral Scientist, 41*(1), 43–63. https://doi.org/10.1177/0002764297041001005

Miller, H. T. (2012). *Governing Narratives: Symbolic Politics and Policy Change*. University of Alabama Press.

Mintzberg, H. (1989). *Mintzberg on Management: Inside Our Strange World of Organizations*. Simon and Schuster.

Moyaert, P. (2007). Can Sublimation be brought about through Idealization? *Ethical Perspectives, 14*(1), 53–78. https://doi.org/10.2143/EP.14.1.2021812

Muhr, S. L., & Kirkegaard, L. (2013). The dream consultant: Productive fantasies at work. *Culture and Organization, 19*(2), 105–123. https://doi.org/10.1080/14759551.2011.644670

Orr, K. (2023). Uncanny organization and the immanence of crisis: The public sector, neoliberalism and Covid-19. *Organization Studies, 44*(12), 2009–2030. https://doi.org/10.1177/01708406231185959

Panizza, F., & Stavrakakis, Y. (2020). Populism, Hegemony, and the Political Construction of "The People": A Discursive Approach. In P. Ostiguy, F. Panizza & B. Moffitt (Eds.), *Populism in Global Perspective* (pp 21–46). Routledge.

Penz, O., & Sauer, B. (2019). *Governing Affects: Neoliberalism, Neo-Bureaucracies, and Service Work*. Routledge.

Petriglieri, G., & Petriglieri, J. L. (2020). The return of the oppressed: A systems psycho-dynamic approach to organization studies. *Academy of Management Annals, 14*(1), 411–449. https://doi.org/10.5465/annals.2017.0007

Phillips, A. (2024). *On Giving Up*. Farrar, Straus and Giroux.

Pierre, J., & Peters, B. G. (2020). *Governance, Politics and the State*. https://public. ebookcentral.proquest.com/choice/publicfullrecord.aspx?p=6234793

Pivnick, B. A., & Hassinger, J. A. (2022). The Relational Citizen as Implicated Subject: Emergent Unconscious Processes in the Psychoanalytic Community Collaboratory. In R. Kabasakalian-McKay & D. Mark (Eds.), *Inhabiting Implication in Racial Oppression and in Relational Psychoanalysis* (pp 158–182). Routledge.

Pohl, L., & Swyngedouw, E. (2023). Enjoying climate change: *Jouissance* as a political factor. *Political Geography, 101*, 102820. https://doi.org/10.1016/j.pol geo.2022.102820

Pottage, A., & Mundy, M. (2004). *Law, Anthropology, and the Constitution of the Social: Making Persons and Things*. Cambridge University Press.

Pressman, J. L., & Wildavsky, A. (1984). *Implementation: How Great Expectations in Washington Are Dashed in Oakland; Or, Why It's Amazing that Federal Programs Work at All, This Being a Saga of the Economic Development Administration as Told by Two Sympathetic Observers Who Seek to Build Morals*. Univ of California Press.

Pusca, A. (2007). Shock, therapy, and postcommunist transitions. *Alternatives, 32*(3), 341–360. https://doi.org/10.1177/030437540703200304

Renström, E. A., Bäck, H., & Carroll, R. (2023). Threats, emotions, and affective polar-ization. *Political Psychology, 44*(6), 1337–1366. https://doi.org/10.1111/pops.12899

Rose, N. (1991). Governing by numbers: Figuring out democracy. *Accounting, Organizations and Society, 16*(7), 673–692. https://doi.org/10.1016/ 0361-3682(91)90019-B

Salberg, J., & Grand, S. (2024). *Transgenerational Trauma: A Contemporary Introduction*. Taylor & Francis.

Savski, K. (2020). Polyphony and polarization in public discourses: Hegemony and dissent in a Slovene policy debate. *Critical Discourse Studies, 17*(4), 377–393. https://doi.org/10.1080/17405904.2019.1609537

Savun, B., & Gineste, C. (2019). From protection to persecution: Threat environment and refugee scapegoating. *Journal of Peace Research, 56*(1), 88–102. https://doi.org/ 10.1177/0022343318811432

Schehr, R. C. (2005). Conventional risk discourse and the proliferation of fear. *Criminal Justice Policy Review, 16*(1), 38–58. https://doi.org/10.1177/0887403404266461

Scherz, P. (2022). *Tomorrow's Troubles: Risk, Anxiety, and Prudence in an Age of Algorithmic Governance*. Georgetown University Press.

Schultz, K., Cattaneo, L. B., Sabina, C., Brunner, L., Jackson, S., & Serrata, J. V. (2016). Key roles of community connectedness in healing from trauma. *Psychology of Violence, 6*(1), 42–48. https://doi.org/10.1037/vio0000025

Scott, J. C. (2020). *Seeing Like a State: How Certain Schemes to Improve the Human Condition Have Failed*. Yale University Press.

Shipley, R., & Michela, J. L. (2006). Can vision motivate planning action? *Planning Practice & Research, 21*(2), 223–244. https://doi.org/10.1080/02697450600944715

Shipley, R., & Newkirk, R. (1998). Visioning: Did anybody see where it came from? *Journal of Planning Literature, 12*(4), 407–416. https://doi.org/10.1177/0885412 29801200402

Thrift, N. (2004). Intensities of feeling: Towards a spatial politics of affect. *Geografiska Annaler: Series B, Human Geography, 86*(1), 57–78. https://doi.org/10.1111/j.0435-3684.2004.00154.x

Toff, B., & Nielsen, R. K. (2022). How news feels: Anticipated anxiety as a factor in news avoidance and a barrier to political engagement. *Political Communication, 39*(6), 697–714. https://doi.org/10.1080/10584609.2022.2123073

Underdal, A. (2010). Complexity and challenges of long-term environmental governance. *Global Environmental Change, 20*(3), 386–393.

Valdre, R. (2019). *On Sublimation: A Path to the Destiny of Desire, Theory, and Treatment*. Routledge. https://doi.org/10.4324/9780429478048

Van Assche, K., Beunen, R., & Duineveld, M. (2012). Performing success and failure in governance: Dutch planning experiences. *Public Administration, 90*(3), 567–581. https://doi.org/10.1111/j.1467-9299.2011.01972.x

Van Assche, K., Beunen, R., & Gruezmacher, M. (2024). *Strategy for Sustainability Transitions: Governance, Community and Environment*. Edward Elgar Publishing.

Van Assche, K., Gruezmacher, M., & Deacon, L. (2020). Land use tools for tempering boom and bust: Strategy and capacity building in governance. *Land Use Policy, 93*, 103994–103994. https://doi.org/10.1016/j.landusepol.2019.05.013

Van Assche, K., Gruezmacher, M., Marais, L., & Perez-Sindin, X. (2023). *Resource Communities: Past Legacies and Future Pathways*. Taylor & Francis. https://books.google.ca/books?id=4EHUEAAAQBAJ

Van Assche, K., Verschraegen, G., & Gruezmacher, M. (2021). Strategy for collectives and common goods: Coordinating strategy, long-term perspectives and policy domains in governance. *Futures, 128*, Article no102716. https://doi.org/10.1016/j.futures.2021.102716

Vanier, A. (2010). Fear, paranoia, and politics. *The Psychoanalytic Review, 97*(2), 215–229. https://doi.org/10.1521/prev.2010.97.2.215

Weinberg, H. (2006). Regression in the group revisited. *Group, 30*(1), 37–53.

Wolf, A., & Schwartz, E. K. (1962). *Psychoanalysis in Groups*. Grune and Stratton.

Zevnik, A. (2017). From fear to anxiety: An exploration into a new socio-political temporality. *Law and Critique, 28*(3), 235–246. https://doi.org/10.1007/s10 978-017-9211-x

Žižek, S. (1992). *Looking Awry: An Introduction to Jacques Lacan through Popular Culture*. MIT Press.

Žižek, S. (2002). *Did Somebody Say Totalitarianism?: Five Interventions in the (mis) use of a Notion*. Verso.

Žižek, S. (2006). *Interrogating the Real*. Bloomsbury Publishing.

Žižek, S. (2019). *The Sublime Object of Ideology*. Verso Books.

Žižek, S. (2020). *The Plague of Fantasies*. Verso Books.

11 Therapy?

Every community limits itself in its own way, and the reasons and mechanisms for this self-limitation are never fully known to the community itself. This is not always a problem, and, if the functioning of the community goes rather smoothly, and demands and conflicts can be handled with relative ease by the routines of the governance system, there might not be a need for a shift in governance, or for an extensive self-analysis. However, problems might arise, unease, anxiety, shocks that find no convincing response, which indicate that something might have to change, and that the future of the community and its governance system are in question. Where self-limitation can be aptly described as self-delusion, in governance or community, or, where in other ways rigidity and blind spots increase, the lack of insight into self, and the resulting mode of self-limitation can cause a wide variety of problems. The implications might become clear only slowly, through a gradual exposure of non-adaptation.

A new perspective might have to be constructed, a new mode of self-observation, which might then make it easier to think of alternative futures. It is possible that distinctions made generations ago, choices institutionalized in the governance of the community when that community was established, need to be rethought. Coming to such insights can be daunting and can require a mode of self-analysis as suggested in this book. Self-analysis is only likely to succeed if a desire for self-analysis is there, and a desire for self-transformation (Fink, 1999). We emphasize the self, because it is the community itself that has to come to insight, and if in the process an insight might be imposed, it is not likely to cause shifts in the self-interpretation, in the feeling and understanding of what is real and meaningful for the community (Fink, 2013a, 2013b).

Communities can come to a reinterpretation of self, of their past and possible and desirable futures. This does not have to take place in governance, and historically, religious revivals, spreading ideologies, burgeoning cities and trade networks, fortuitous meetings, and unfortunate adventures have all contributed to shifts in both thinking and organization that can add up to the

DOI: 10.4324/9781032696676-11

emergence of new communities and the transformation of others (Anderson et al., 2020; Boin et al., 2013). Yet, at some point, governance structures emerge, and, in democratic polities, new self-understandings will assert themselves there (North et al., 2009), while governance itself can contribute to a more structured reflection on, and process of self-transformation.

Governance can be the site of emergence for new narratives, or the site of consideration, of a comparison of narratives old and new. It can be the site to grasp anew what is of value for the community, and which values mark the community, and this can engender a reflection on the past that produced those assets, and on an environment where those assets and those cultural values might be less prized, even problematic (Antze & Lambek, 2016). Psychoanalytic perspectives on governance can, in our view, offer great promise of deeper self-understanding, where communities are becoming aware that change is needed, but are not yet able to sense where the issue might be. Self-analysis will work better when communities or their governance systems do not limit the set of potential problems to a set of technical issues (in governance) or to stark ideological choices, known in advance (Allmendinger, 2017; Hillier, 2003). Put more strongly, it can only work when there is a genuine desire to find out where current self-understandings are stuck, which problems they cause, and how the genesis of these problems can be located in the governance path and the history of the embedding community.

Community and identity

Individuals become unique individuals and can function because they find definitional and operational support with others, in communities small and large, starting from the family. Communities cannot exist without individuals and the other way around. Luhmann spoke of emergent co-evolution, of psychic and social systems (Luhmann, 1995). For Luhmann, for the psychoanalytic tradition, and for us, meaning is the shared medium in this co-evolution, and meaning operates through narrative. The meanings we bestow upon the world are not qualitatively different from the meanings we give to ourselves (Yanow, 2000; Gunder & Hillier, 2016). We understand ourselves and the world through narrative, while the feelings we have about ourselves and the world borrow from the same shared vocabulary and narrative encyclopedia.

Communities of some sort are necessary for individuals to grow up and flourish, yet not all sorts of communities we are familiar with nowadays were historically present, and therefore necessary. Nation-states did not need to exist for humans to flourish (Magris, 2016; Žižek, 2009). Whatever political entity exists, does not have to be the source of identification for the people living in its territory (Luhmann, 1987; Marks et al., 2009). Noncontiguous and non-territorial communities and polities existed and exist; communities do not necessarily govern themselves. Community identity, in other words, does not have to be key to governance, and community futures are not always envisioned through governance.

What we proposed in this book, inspired by both Lacanian and Freudian psychoanalysis, is that community identity is not always there. As an Ego function, it can appear and disappear as an effect of communication, and possibly reinforced through governance, where the community must be represented and reflected upon. For us, narrative identity can stabilize itself, and amplify its effects on community life, through institutionalization, through achieving a grip on and developing the structures of governance (Miller, 2002; Van Assche et al., 2024). If a group identifies through language as a symbol of unity, and that group can organize education, preferring that language, the group can reproduce, increase, and possibly build its status. Narrative identity is in the Lacanian order of the imaginary, and it can, but does not have to engender Voice in community governance, while the existence of governance structures can encourage the formation and amplification of Voice.

Not every polity, every governance system, is therefore associated with a narrative identity, yet if communities organize, if governance crystallizes, this will be a forceful support for the emergence of Voice, while a strong community Ego will support the development of governance systems. Internal diversity can be acknowledged or not through governance, and images of self and environment can be created that now become amenable to collective scrutiny and to the taking of collectively binding decisions and the articulation of guiding institutions. Stories about past and present, about self and environment, make it possible to construct stories about the future, to articulate collective sentiment and expectation (Glynos, 2011; Žižek, 2019).

A second identity exists alongside the collective Ego, a more persistent one, one that we call the *autopoietic identity*, following Luhmann, an identity which expresses the unity and unicity of the self-reproduction of the governance system. Each governance system is autopoietic, and is marked by an autopoietic identity, a unique interplay of the actor/institution and power/knowledge configuration, a unique way of relating to its environment (Beunen et al., 2016; Luhmann, 1990). Autopoietic identities thus belong to governance, while narrative identity, or collective Ego, can come into being before governance appears, but is likely to stabilize and materialize itself through governance.

In Lacanian terms, autopoietic identity has both Real and Symbolic dimensions, as its functioning is shaped by exteriority as well as interiority, and as it explains itself as based on a law it imposed on itself (Eyers, 2012). Communities are not fully aware of the autopoietic identity of their governance system, and, hence of themselves, as governance will guide the self-reproduction of the community (Luhmann, 1997). Actors in governance themselves are not fully aware of the nature of the governance path, of all aspects of self-transformation, all drivers and mechanisms at work. Other features are more accessible to observation, and to change (Valentinov et al., 2019).

Both the autopoietic and the narrative identity of a community have conscious and unconscious aspects (Freud, 2003). Communities can have desires and emotions, mediated through narrative; they share values, ideas, and histories that give meaning to themselves and their world. Individuals can orient themselves through but also in distinction with such discursive features. Autopoietic identities can include traces of old narrative identities, while narrative identities can refer to, but can also be shaped by autopoietic identities. Communities forget things, do not want to know things, feel things, and some of this will be pushed into a collective unconscious, where it can still form rhizomatic, associative connections with other ideas, feelings, narratives, histories, and objects (Freud, 2015). Repression is possible as a collective and for individuals, with collective repression validating and enabling individual repression.

Collective desire, collective fears and hopes do not disappear by decree, and the writing of new rules, the production of new institutions in governance, even if sensitive to the mood in the community, will only imperfectly respond to it (Grint & Jones, 2022; Holstein & Gubrium, 1999). Moreover, as the community is imperfectly aware of its own desires, as the process of governance entails a series of reinterpretations of collective desire and its big concepts, and as desire moves in unpredictable ways, governance will always disappoint, which is positively a driver of continuous change (Hirvonen, 2017; John et al., 2015).

The collective subject, for us, comprises conscious and unconscious functions and it represents the unity of the autopoietic and narrative identity. The subject does not always come into being. It is the result of the production of interiority that integrated elements from a series of external contexts, and it is a product of both thinking and organizing. It is, furthermore, a split subject, not transparent to itself and reliant on shifting identifications to stabilize itself. It is compelled to exaggerate its cohesion and its certainty by means of productive fictions, selective blindness, and various techniques of de-paradoxification. It is a precarious product of self-organization that is nevertheless responsible for most aspects of the way we organize and define ourselves (McGowan, 2012; Vanheule, 2016).

Governance faces paradoxes because it cannot acknowledge this absence of solid ground, the facts of contingency and self-organization (Beunen et al., 2016). It can never be sure whether it must reflect what exists in the community or must shape that community. The real dilemma is not a moral or legal one but an ontological one already encountered. Governance cannot count on something existing out there to give it direction; it cannot really explain its own existence based on what is out there and it cannot be sure how to shape the community (Kapoor et al., 2023; Walker & Shove, 2007). Neither alternative, neither direction of creation has any ultimate grounding.

Structuring of therapy?

The psychoanalytic understanding of governance proposed in this book cannot undo the contingency of governance and community, and offer certainties about good governance, the real community, about absolute values. Those certainties do not exist, although sometimes it is helpful for leadership to pretend they do (Swyngedouw, 2022; Vaara et al., 2016). What this perspective can offer, is an insight into the productive sides of this contingency, into the possibilities for reinterpretation it offers (Pottage & Sherman, 2010). Such reinterpretation of self comes through a reinterpretation of history, and needs to be a self-interpretation, emerging, likely slowly, from a self-analysis. Given the pressures on governance, the pressures on time and resources, and the difficulties in shifting patterns of participation and representation, a slow self-analysis is not likely to be proposed by those playing a circumscribed role in governance. It will have to be proposed by leadership, or by influential outsiders.

It will most likely be accepted where uncertainty about the path forward is already creeping in, when doubts are spreading about the direction the community is going. Even then, in many cases the problem will be positioned automatically in a corner identified by current ideologies and Master Signifiers (Stavrakakis, 2002; Žižek, 2020). Routine blaming can even reinforce local identity, yet this does not sharpen the analysis of actual problems. What comes into the picture when recognizing problems is identified according to the existing checklist of problems, to be solved according to the current menu of solutions (Brans & Rossbach, 1997; Jun, 2012). Routines in thinking and organizing, both unexamined, make for such reproduction of the system, even when trying to fix it (Beunen & Patterson, 2017).

Self-analysis can become accessible through the analysis of one particular problem, one particular policy domain (maybe requiring a review), or, when a more comprehensive visioning process is in the cards, when an agreed-upon examination of desirable futures can make an examination of current desires more palatable. What *should* be the entry point for a prolonged self-reflection cannot be decided a priori or postulated in general. *When* a community is ready to think about itself in a different way, to question some of its certainties, to rethink its own history, can only be ascertained by people in the community. What can be the right entrance also remains invisible from the outside. Imposing sustainability planning on barely sustainable rural communities is not going to lead to much reflection or sustainability; rather, it will lead to an intensification of pre-coded negative feelings toward those imposing such a wasteful exercise (Gunder & Hillier, 2009; Hulme, 2009).

Intensified self-analysis can start from path mapping (see Chapter 8), or it can start from a problem, which then brings about a limited form of path mapping. It can start with an examination of assumptions about identity, a slow exploration of narrative identity, and from there, through a mapping of legacies, to autopoietic identity. Or, an affective angle becomes possible, where positive and negative feelings about the community, its future and past, are explored, and an interrogation of those feelings becomes possible, a connection to past events, to key narratives, Master Signifiers, and desires. Speaking truly freely (see earlier remarks on visioning) and truly associatively is hard in a group setting, yet, if there is a sense of shared purpose, of the seriousness of the problem, and a willingness to question local certainties, part of the defenses might be overcome and group pressure might also work positively in this effort (Scheidlinger, 1968; Wolf & Schwartz, 1962).

Small groups in key positions, possibly crosscutting leadership groups, or groups assembled by leadership can take the lead, can open the process, and set the example (Boin et al., 2013; Casey et al., 2017). Smaller groups with more insight into community and governance, that are more aware of the issues and the limited problem-solving capacities of the community, can offer opportunities for experimentation, for a probing of the willingness to investigate where things got stuck, where the community is trapped in constraining forms of thinking and organizing. Such small groups might also identify what could work as a focus for a more participatory, larger-scale self-examination. Commonly, both problems and solutions are positioned externally, and in the realm of missing resources, infrastructures, and powers (Alvesson & Spicer, 2010). Most likely, there is truth to these assessments, but also an investment in them that has other sources, which need to be identified before alternative solutions and different problem definitions become possible (Abse & Jessner, 2019; Glynos & Stavrakakis, 2008).

The mapping of histories, of identity narratives, of hopes and fears about the future can combine with the mapping of the governance path, with public discussion, and visioning sessions, to trace the formation of community identities, and the role of governance in shaping and reflecting them. This cannot be an encyclopedic exercise, or an objective one, as the point is not to determine all the facts but to identify, in governance paths, where modes of thinking and organizing came in and got entrenched, how stories became naturalized, and where futures were abandoned or set in stone. Becoming aware of false certainties opens avenues for exploring alternative versions of reality, and alternative futures. The concepts offered in all the chapters of this book can be helpful in such endeavor, and for each process a different selection of concepts will make sense and can be more easily translated into method. One can imagine a careful mapping of patterns of inclusion and exclusion in one self-analysis, an interrogation of Master Signifiers elsewhere, or an intense self-questioning regarding ambition levels and policy direction in another. The unearthing of ambivalence and incoherence can offer openings, or the patient consideration of anxieties surrounding one version of history, one policy domain, one key decision (cf. Czarniawska, 2014; Glynos, 2011; Miller, 2002; Yanow, 2000).

While the entry points, in some cases the excuse for embarking on a self-analysis, can be quite diverse, the analysis can thus also take a variety of formats, in which the process can unfold in even more ways.

Forms of mapping can give space to such unfolding, as just indicated. Yet, starting with the future, a process of visioning remains possible. Our earlier paragraphs on visioning argued that, ideally, self-analysis precedes visioning, so the visioning can more freely explore different desirable futures and recognize more easily how realistic different scenarios might be. Here we add that sometimes visioning might be the only starting point available to interrogate the past. Considering the future is less likely to be seen a waste of time, might even be mandated every few years, and might offer the opportunity to ask difficult questions, starting with the obvious ones: How did we get here? Why is it so difficult to move forward for us?

A third type of self-analysis asks directly, what could be self-limiting? One can start here with eliciting, by listening to the technical answers, the expected answers, the culturally correct answers, the safe answers, and work from there. Blind spots can be traced, limiting understandings of governance, good governance, the real job, the good life, ideas on development, and of core values and activities. Master Signifiers can be identified, not-so-productive fictions recognized, and unrecognized desires, latent affects, and, hopefully, from there, the reasons for their force and their intransigence can be determined. Sometimes, such a path might lead to the return of repressed signifiers, elsewhere to the dispelling of modernist or neoliberal mythologies of governance and its limited functions.

A start and an end?

While a uniform structure of self-analysis cannot be expected or imposed, guiding principles can be formulated, and the same can be said for the beginning and end (cf Van Assche et al., 2023). What we regard as therapy is not always possible and necessary. It must be desired and accepted by the community. This desire can start with a smaller group, which might be able to convince a larger group, and the therapy needs to be conceptualized as an intensified period of self-reflection. To avoid jumping to conclusions about self and future, it might be good to insert a period of unwinding after intense self-reflection, for the dust to settle, and then go to work, if so desired, on outlining the policy implications of the insights gained. As stated, a reflection on policy can be a starting point, but if there is a real analysis triggered by it, then a breather afterwards, and a reconsideration of policy implications are still warranted.

An intensified period of self-reflection can help disrupt problematic patterns of self-reproducing discourse and affect. It can expose knots in thinking and organizing that in effect undermine self-reflection, immunize the community for critique and self-critique, and thus work as self-limiting and nonadaptive devices (Van Assche et al., 2022). Adaptation does not represent an ideal of perfect response to the requirements of an objective reality. Adaptation here is only used to point at the dangers of multiplying blind spots, in terms of risks, in terms of internal exclusions and injustice, and in terms of opportunities for self-transformation that might become hard to imagine and hard to organize.

Bringing back more diverse perspectives in governance, in the relation between governance and community, allowing for and designing thoughtful redundancies in the governance system, safeguarding checks and balances,

increasing governance capacity, remediating unfortunate exclusions, culti-
vating reflexivity and second-order observation are all goals that can appear
and be specified and contextualized *after* self-analysis, and can even also be
presented as positive values and features more generally. Reform of governance can be proposed as a way to lay the groundwork for a next round of
visioning and strategizing, where substance can be given to the desired future
of the community (Van Assche et al., 2020). Once again, the order, selection, intensity, and implications of the elements in the suite of self-analysis,
visioning, and strategy will have to be decided in and by the community, with
the main consideration probably that actual analysis is not avoided. Pressures
will likely mount to do something useful, something real, something understandable, and the patience, trust, and openness needed for self-analysis to
work are scarce goods (Demertzis, 2019). Hence, the importance of leadership, for guiding and shaping self-analysis, and for deciding on a starting
point, temporally and conceptually. We come back to this in the next section.

We note that some problems make therapy more necessary yet harder to
propose. Histories of trauma, of oppression, and of fragmentation and weak
governance leave legacies in thinking and organizing that are hard to spot
from the point of view of most community members (Giesen & Eisenstadt,
2015; Mohatt et al., 2014). Politics and administration cannot diverge too
much from the community perspective. People in politics and administration
might have capitalized on the forms of closure, the grievances and hardened
realities stemming from such difficult histories to get elected, to build a
career. External actors, higher-level administrative and political actors, but
also multinational companies and sometimes large nonprofit organizations
might have played a long game of divide and conquer, leaving communities
weaker and polarized (North et al., 2009; Vélez-Torres et al., 2022).

Where forms of governance and forms of leadership are anchored in a transcendental order, self-questioning can be out of the question (Žižek, 2009).
If a leader is appointed by God, reigns by divine right, and if rules or even
governance structures are dictated directly by a higher power, then standing
up to leadership, rethinking governance is not an easy choice. It amounts to
defying God, or at least, misrecognizing the order created by him (Collinson,
2011; Hook, 2011). Yet, divinely inspired ideologies do not stop drives and
desires, do not erase the unconscious, and do not prevent problematic legacies from developing. Making the unconscious conscious, widening the repertoire of adaptations and reinventions will be a tall order, and what we can
empirically observe is a likelihood of either revolution or a slow erosion of
the transcendental semantics (Hoggett, 2006; Matthies-Boon, 2017). If this
occurs, the question is whether leadership also believes it does not have to
operate on this set of beliefs anymore. People in such places tend to vote with
their feet, escape the regime, or less dramatically, move away from the village
and pursue studies and career somewhere else.

As we are speaking about starting points for analysis, we need to point at
several structural possibilities of analysis coming after what ideally would

be the result of analysis. We mentioned the use of visioning as an introduction to self-analysis, and we add here that transitional forms of governance can also feature both as a result of and as a preface to analysis. Transitional forms of governance (see Van Assche et al., 2023) can emerge where routines were not functional anymore, and they can be consciously designed as temporary configurations, enabling the building of governance capacity. Where futures cannot be easily envisioned or organized, or, where something has to be done immediately, transitional forms of governance can offer respite, can either save the day or make it possible to organize for tomorrow. The nature of transitional governance can thus either enable long-term governance or remain in the present, and it comes with risks of entrenching itself, as interim leaders and councils might have the inclination to stay.

A transitional form of governance can also be built *before* analysis, not as an outcome of analysis, in the knowledge that it will create conditions for strategy, but as something presented as useful without reference to self-analysis or strategy. Leadership will have to take the initiative and take a risk. One can raise ethical questions, and leadership might be punished afterward. Yet, for the greater good, it might be worth taking the risk of creating conditions where the community can choose more freely whether to embark on a self-analysis or not. If a community is attached to its previous routines, ideologies, and other fantasies, then transitional governance might lead to a seeming return of the old system, even when the screen of fantasy must cover the differences between old and new. If a community has lost its desire to rethink itself or to go back to the past, a quiet transformation of the temporary regime into a permanent one might be on the horizon. If, however, leadership is able temporarily to shift patterns of decision-making and increase the capacity of the community to reflect, then transitional governance can be helpful for self-analysis and for strategizing. Then, transitional governance can be a starting point for self-analysis, or for exploring alternative futures.

If transitional governance precedes self-analysis, we are speaking of a truly temporary form of governance, a phase of either quickly fixing urgent problems or slowly opening discussions, asking questions, to bring the community to a point where it could, if it desires to do so, choose a more intensive self-analysis. Leadership, civil society organizations, media, higher-level actors, and external experts can all play a role, but if we are speaking of democracies, the process cannot be fully coordinated. These are not the conditions for a formal redesign of the governance system, nor for an orchestrated campaign to question the certainties the community is living with. Transitional governance, if it precedes self-analysis because it is an inheritance of shocks and instabilities in the community, is more problematic, as it requires more obviously an analysis of instabilities and inheritance, and as it cannot easily function as a platform for envisioning the future. A similarity with other situations where thorough self-analysis might follow transitional governance, however, is that also here, a nudging toward reflexivity and diversity in perspectives, in governance and community, is in place. People need to be truly interested in exploring alternative futures, and in rethinking themselves. This, not strangely, might take some persuading, and where transitional governance needs to come in first, laying the groundwork for public and governance discourse to open to self-reflection and consideration of futures, has to be at the core of leadership efforts. An alternative is to bestow more power on leadership, temporarily, and hope for them to take decisions that can initiate a collective self-reflection later. Pandemics and disasters are common situations for such a transfer of power.

If the community then decides that a form of self-analysis is desirable, we can refer to the previous section and previous chapters, for inspiration. Right now, we will look at the question on how to *end* the analysis, a prickly topic in the history of psychoanalysis, where Freud himself changed his mind a few times (Sandler, 1991). Self-analysis must be a phase of exception, a more participatory phase, where both stories and modes of organization are slowly brought into question. Such intense self-scrutiny cannot be sustained (see also Chapter 3), and routines must be reinstalled, albeit maybe different ones. Intense participation, intense self-reflection, questioning of the fabric of the community, which might or not be a home and a source of identification for most people, none of this can be sustained forever (Cooke & Kothari, 2001). In some cases, the analysis will point more at changes in organizing, elsewhere to a need for rethinking stories, while a clearer understanding of which desires might really be given a place in the life of the community is always at stake.

Another reason why interminable therapy might not be workable, is that governance and community might unravel in the process itself, that neither short- nor long-term decisions are taken or coordinated. The community can fall apart; anarchy can ensue. In practice, this is not too likely, as various legal, political, cultural, and financial stops are built into democratic governance systems, and few would see benefit in sowing chaos (Luhmann, 2018). Except, that is, when democratic principles are not valued, or believed to function locally, and where weak governance brings factional gains (Acemoglu & Robinson, 2013). A related but distinct risk is that leadership might concentrate too much power in its hands and use the intention of self-analysis to drag out the process indefinitely and get rid of democratic checks (Tourish, 2013). This scenario, too, seems unlikely, as the conditions that might enable self-analysis are very improbably the conditions which would allow for such authoritarian takeover of a transitional state. More subtle, and more realistic, is a creeping dependence on leadership guiding a self-analysis (see below), which might translate into a lack of autonomy and judgment afterward.

In each case, an end to therapy needs to be envisioned and agreed upon, even if not all are convinced of the results. In governance, this can mean the end of a period of intensified participation, of community creation through increased and diversified forms of social and cultural organization that can link to the reflection on past and future in the governance system (Minkler, 2012). One can also say that the governance system itself is temporarily extended, and that the boundary between governance and community is temporarily blurred, since the separation of a subsystem guiding the functioning of the community, in the image of an ideal Ego, might not work for the time being.

An ending to our proposed version of community therapy must be demarcated, possibly ritually, and also in a manner recognizable for the governance system itself. Governance needs to be restarted according to clear rules, and reflexivity must be given a clear and more limited space again. It does require a space, as lack of reflexivity and second-order observation, in combination with lack of diversity in perspective, was probably part of the problem beforehand (Van Assche et al., 2024). Roles must be taken up again, people must go back to their main job, even if that job might be redefined. Next steps need to be agreed upon. This might mean that no strategy is deemed possible yet, but steps in governance reform are agreed upon. Or, a clear process for strategy formulation can be envisioned, with moments of intense participation, but without reopening all discussions of the self-analysis. Elsewhere, strategizing can be included in, or even be the starting point for self-analysis. In such a situation, what is the end of therapy is agreement on the final form of the strategy. Such a path might be more likely for smaller communities, as it might destabilize intricate patterns of role differentiation in larger ones; there, self-analysis through subsystems is more likely to work.

In self-analysis, a community might have to traverse its grounding fantasy, which can be traumatic itself, casting the community in an unsettling light, and making previous forms of enjoyment appear as problematic (Bistoen et al., 2014; McGowan, 2012). This can easily trigger resentment, and the reformation of community in a manner that cements new exclusions and settles around new Master Signifiers felt as old and authentic. The process of analysis itself can be too burdensome, and become associated with despised ideologies, a recognition that can then crystallize a counter-ideology or identity entirely at odds with the likely intention of softening the hold of ideology, of old stories and identities, of unproductive fictions over the community and its governance (Renström et al., 2023).

When such entrenching backlash does not take place, when the need for stability, identity, unity, and simplicity does not come to outweigh the desire for unveiling what keeps us trapped in unworkable versions of past, present, and future, a new phase of community formation can take place. One can speak of a phase of restarting Ego formation, of reformulation and reforming the community, in a self-analysis where the line between community and governance is temporarily blurred, a blurring that can enable a reorientation of thinking and organizing (Czarniawska & Mazza, 2003). After this, a new separation of community and governance is required, a reinstalling of roles and routines, or, in systems terms, the re-separating of the subsystem that can be entrusted with steering and further self-transformation (Luhmann et al., 2004; Seidl, 2016). This might seem the utopian inverse of the gloomy extreme of community dissolution through self-doubt just mentioned.

One can counterpose that one must be careful with normalizing ideas of stability, unity, and identity in governance. In other words, in assessing what is utopian or dystopian, what is desirable and what is not, one cannot take for granted the mythologies communities tell about themselves and their governance systems (Žižek, 2009). Nor can one uncritically accept the essentializing models of democracy, of politics and administration dominating the literature (Miller, 2002). If we understand communities as inherently split because of governance, but also as coming into being through governance, and if we take co-evolution seriously, one cannot assume that a stable essence of democracy exists, nor an associated stable definition of "good governance" (Elwood, 2004; Swyngedouw, 2022). Nor can one presume that stability, unity, and Voice are the norm. Narratives about stability, as much as narratives about identity and cohesion serve to further stabilize; they are not a sign of stability (Stavrakakis, 2010). They are part of the repertoire of devices of de-paradoxification. In practice, what is strived for is rare, and tension and instability are the norm. They are also what is needed to keep governance and community evolving.

For an impartial outsider not paying attention to the political science literature, or to the self-descriptions of communities, what is most common is a stunning variety of situations and phases (Miller, 2012; Vaara et al., 2016). Even looking at a single community, one could observe that sometimes it is a community, sometimes not; sometimes it is able to represent itself in governance, yet not always. Voice can emerge, can be able to reshape the community according to new Master Signifiers and newly discovered desires. A conflict can erupt, productively, a conflict that could forge a new community where fragmentation existed. Yet, in the same community, nearly imperceptible differences can in time lead to the recognition of incompatible *objets petit a*, leading thus to irresolvable polarization. Meta-narratives of good governance, of the role identity ought to play in governance, evolve with the community (Rasche & Seidl, 2020). Fictions of certainty and necessity, of one mode of organization, one possible future are most likely already questioned in corners of the community (Scott, 1998).

The seemingly radical stories of community formation and community dissolution through intense self-analysis are thus less improbable than it might seem. One has to bear in mind that communities do not have the stability of individuals, and that even for individuals, stories of stable and unified identity are still stories, fictions that can be productive and sometimes less so (Holstein & Gubrium, 1999).

Leaders and therapists

If there is therapy, is there a therapist? Yes, but not necessarily as an expert for hire. We distinguish situations where leadership can play the role of therapist, and situations where leadership is involved, but assisted by experts. Those experts can be outsiders, but they can also be trusted community members, maybe elders, maybe people who are deeply familiar with the community, who have broad networks and varied experiences.

Closed communities, continuously referring to a stereotyped past, to tradition and identity, might be exceptionally sensitive to outsiders offering knowledge or perspective that seem to question values and identities, to interrogate local certainties, processes of naturalization, and to suggest latent drivers of policy (Van Assche et al., 2023). Insiders are thus preferable, but those insiders take a risk, and might lose their insider status and their networks and support systems by offering the questions and proposing a period of self-questioning. Leaders can take on the role of therapist, or ally of the therapist, or assembler of a smaller collective that can play a leading role in the self-analysis, a role straddling insider and outsider perspectives (Jarrett & Vince, 2023).

Leadership might have been aware of many of the issues unspoken in the community, might have experience elsewhere or connections, and might have other modes of reality testing for interpretations of the present and for visions for the future. The risk for leadership remains, however, and the actual position as semi-outsider cannot be transformed into a public perception as outsider, and a political outsider status. Leadership, however, can be understood in a distributed manner (Bolden et al., 2023), and new sites of leadership can be created so both the emotional burden and social risk can be distributed, away from the shoulders of formal leadership (Bush, 2023).

Not all the tools of the therapist translate well from the individual to the community. Transference, as a driver of self-exploration, by projection of unconscious roles, relations, affects, fantasies, and desires onto a therapist, who can then assess how to follow with questions intended to accompany and guide associations, how to explore the unconscious signifiers shape current behavior (Fink, 2017b), cannot be pursued consciously by whoever is structuring the self-reflection. Leadership or trusted insiders or outsiders assisting them, or the other way around, with others being the front person, with the backing of leadership, do not have the training to elicit and manage transference. Bringing in a consultant with psychoanalytic background, or, more plausibly, a counseling psychologist with experience in working with organizations, or maybe experience with mediation, will not likely provide an answer as a knowledge of governance and community cannot be missed. A strategy consultant, most often with a business background, often comes with preformed ideas on strategy, and tends to be less sensitive to the affective and narrative dimensions of governance, hence to the need for self-analysis in these dimensions (Jarzabkowski & Seidl, 2008; Shipley & Utz, 2012).

While we cannot count on wise guidance of the process of transference in community self-analysis, one can still expect it to occur, and one cannot dismiss it (Moxnes, 1998; Wolf & Schwartz, 1962). One very basic principle is that small teams of people with different skills and different positions on the insider/outsider spectrum, are most likely to be effective. A second, still very basic principle, for those guiding the analysis, would be that they cannot take anything personally (cf. Fink, 2017a). Whatever emerges could refer to other people and things in the past. The more freely people can speak, the more power differences and the need to act or follow everyday roles moves to the background, the bigger chance that actual (conscious) opinion shows up. Hopefully, in a moment when guards are down and participants trust nothing will be held against them, the unconscious can come to speak (McSwite, 1997). When others echo feelings and stories coming to the surface, this can signify something in the collective unconscious (Rutan, 1999). Those who guide the analysis can predict some of this, through familiarity with the community, through participation in governance, through knowledge of public discourse and the assumptions, affects, and latent signifiers at work there. For the channeling of transference, a stable, small, and modest guiding team might be advisable. Stability and an ability to remain in the background, to avoid prickly statements and all too guiding questions can create a team akin to the blank slate of classic psychoanalysis, a canvas for transference, on which feelings can be projected. This only works, we posit, if the guiding team is not overly hierarchical, and if the focus of the meetings is not overtly focused on dialectical learning, where prickliness does have its place.

The guides, or therapists, cannot encourage entirely free association, as this would not fit the image of the exercise for the participants, yet it is possible to pick up on signals, on affects, pauses, on unusual signifiers, on cracks or internal contradictions in the discourse, on silences, and on seemingly illogical associations between the content, form, and emotions of the story. It also remains possible to meander and find apologies, within the framing narrative of the process (as visioning or otherwise) for such meandering, for following sidetracks, stories that might expose affects belonging to stories disavowed publicly or even privately. Small questions can encourage the exploration of such little sidetracks, which seem to go nowhere but might expose what is unsaid (Sabbadini, 1992).

The process can take place in several circles, in and outside governance, where insights from previous series of sessions, possibly more private, possibly within leadership circles themselves, can serve as input for the next phase. The transition between phases needs to be carefully calibrated, as the insights a rather audacious leadership circle is coming to, might not all be effective when publicized or when used as starting point for more public sessions. There might be a delicate balancing between sharing and moving things faster, and on the other hand, allowing people to discover things for themselves (McCann, 2001; Misztal, 2003). Moreover, there is a difference, despite the bracketing of roles and power in self-analysis, between the roles people must go back to. We refer here to power relations that cannot be erased, whatever the situation, and we refer to a difference between those who must play a formal role in governance after self-analysis, and those who are there to speak, hopefully listen and explore, but who have no responsibility later (Elwood, 2004).

We can shift the perspective slightly and consider which roles leadership can play in communities where the future is unclear or threatening, where it is not apparent what would count as an asset toward community development, or where the present is haunted by a past that is only half understood (Gunn & Hillier, 2014). Often, we know now, both things happen at the same time, and such communities would certainly benefit from self-analysis in the manner proposed. We suggest here to distinguish four leadership roles, which can occur in a distributed fashion in such situations, that can all be helpful toward self-analysis and, where this is possible, toward visioning and strategizing. We speak of leaders as *builders, translators, brokers,* and, finally, *therapists.* Not all roles can be combined in one person, and not all will be practicable and desirable for each community.

Leaders as builders are not only welcome where ambitions are high, but also where the base to build from is wobbly. Building governance capacity or building community can take precedence, or both can be pursued at the same time (Minkler, 2012). A group can become a community through development of governance, or in different ways, while a community can stabilize and transform itself through governance. As noted often enough, the split between governance and community will remain and introduce instability, which can be productive as well as problematic. The organization of governance, the crafting of stories of shared values, pasts, and futures; the development and strategic use of public discourse; and the installation of specialized roles and institutions in governance can all contribute to a co-evolution of governance and community that can stabilize both and enhance the capacity of the community to steer itself.

Leaders as brokers can build networks inside and outside the community, between governance and community. They can find common ground that can stabilize relations and make those relations instrumental for the community and/or its governance capacity. Brokers can operate strategically, or less so, but even where conscious strategy is absent, unconscious desire and shared values can be at work, making brokerage successful (Burt et al., 2021). Sometimes, connections are waiting to be made; the potential is there for pieces to be put together and reinforce modes of thinking and organizing that benefit governance and community. Sometimes, simple co-presence leads to the formation of shared stories. Elsewhere, a narrative identity, or a latent vision for the future can bring one to establish connections with people, organizations, and communities not noticed before (Shields, 2005).

Leaders as brokers are often, but not necessarily, translators. The finding of common ground often involves the finding of a common language, or the creative translation of one language to another. Different groups might be barred from cooperation, coordination, or even communication by discourses that seem alien to each other (Alvesson & Spicer, 2010). Yet, creative translation might be able to reduce this alienation, by identifying similarities between Master Signifiers, by working with or creating shared desires, perhaps first identifying unacknowledged desires existing with both (Helsloot & Groenendaal, 2017). Shared experiences, positive and negative, can bring groups closer together, but so can new stories about old experiences, which place them in a different light and smoothen the finding of connections.

Leaders as therapists are willing and able to unravel, maybe with others, trusted insiders or outsiders, the patterns of thinking and organizing, conscious and unconscious, that create symptoms, that keep communities trapped in versions of the present which make thinking and organizing for alternative futures more difficult. Leaders as therapists are willing to take a distance in this aspect of their role, from the day-to-day preoccupations, to think about the versions and legacies of the past that render it difficult to see what is possible and desirable for the future, and to choose more freely for one or another imagined future (Van Assche et al., 2023). This means that the power of governance and the power of discourse and desire to naturalize what is contingent has to be grasped, and harnessed where necessary (Hoggett, 2015).

Even where leaders function as therapists, self-analysis as proposed here is not always warranted, and good therapists do not sell therapy as the order of the day. Insight gained by them can inspire subtle governance reform, or measured interventions in public discourse. It can inspire their activities as broker, builder, and translator, and other functions. As suggested, more intense forms of self-analysis might become possible in smaller leadership teams, or where competition between leadership (factions) is too high. Small networks of promising people can be formed, possibly complemented by a small group of external advisers and allies, networks able to intensify self-reflection and to think more easily beyond current roles and realities.

The trio of self-analysis, visioning, and strategy where it is deemed possible and desirable, where self-analysis therefore has the furthest reaching implications, implies that all four roles identified above should come into action. Likely, the emphasis shifts in the process, and different people might come to the foreground over time. Ideally, leadership is willing to recognize qualities useful for the stages of the process, and open to bringing new people into its ranks, or new advisers who could play a role in self-analysis, visioning, or strategy. Many specialties in advisory consulting exist, which do not need to be listed here, yet, as always, we would warn against taking titles and promises at face value since, for sensitive roles in reinvention processes, leadership ought to be very well acquainted with the advisory product offered. In the other direction, familiarity with the community is not always crucial. It depends on the role taken and phase of the process involved in (e.g., self-analysis vs. strategy), on the difficulty of the situation at hand (already a clear understanding of issues, or not at all?), on the willingness to learn about process and community, and on other factors such as personality and cultural familiarity.

A note on governance paths

A reconstruction, a careful mapping of governance paths can be helpful in the process of self-analysis, but also for visioning and strategizing. Who is doing this, how participatory the process might be, how comprehensive, what needs to be focused upon, can only be determined in the situation (Van Assche et al., 2019). We mentioned before that it will be of little value if it is seen merely as information, either encyclopedic or as a sketch.

As we are nearing the close of this book, it makes sense to emphasize that the framing of the co-evolution of governance and community through the method of path mapping might look cumbersome, but we argue it offers a flexible heuristic in which all the concepts discussed in this book could be given a place. Each community, each process will make some concepts more relevant than others since problems and qualities are different, since what is rigid and flexible will be delineated in different ways in each governance path. Analysis of governance paths can help to grasp how ideas, actors, and stories entered governance; how futures became imaginable and amenable to strategic organization; and how versions of the past became present or left legacies in governance. Reconstruction of governance paths can assist in discerning the modes of self-transformation at work, formally and informally, and potentially available. It can show the weaknesses in governance capacity, where they came from, how they might associate with dominant narratives, with collective affect, with shocks or traumatic events.

We note that there is no need to structure the whole process around the reconstruction of governance paths, and that, in fact, self-analysis can proceed in the other direction, from the present to the past. Such slow elicitation of stories, affects, and events related to the past, can aim at the reconstruction of governance paths, not for its own sake, but to elucidate what shaped the current forms of self-delimitation. One can proceed like this without the aim of coming to an image of the governance path. Whatever path is chosen, it does remain helpful to have a general framing of governance evolution at hand, either for use by all involved, or for those who are guiding it, a framing that might be the result of a previous phase of self-analysis in smaller circle.

Concluding: self-limitation

This book, its perspective on governance and community and their always unstable relations that are nevertheless necessary to stabilize both, can be used to inspire leadership and community toward self-analysis and strategy, toward overcoming self-limitation and toward the discovery of more rewarding futures. We discussed the favorable conditions for self-analysis, inspired by psychoanalysis, and its place in relation to visioning and strategy, in the previous sections.

Not every community is interested and able to embark on the form of self-analysis we called community therapy, an intense form of self-reflection inspired by both psychoanalysis and governance theory. Not every community needs it, as problematic forms of self-delimitation, stemming from entwined traditions of thinking and organizing, might not be an issue. If there are such problems, they do not always require therapy as they might be dispelled by the usual political and administrative processes. Where lections and administrative routines, but also administrative reforms requiring less self-reflection can resolve the issue, there might be no need to consider the proposed intense form of self-analysis.

A different type of situation, which is not uncommon, appears where therapeutic leadership, which might be concentrated in governance or outside, is able to see beyond the realities of the day and mobilize other leadership functions, to move the community in a different direction, beyond current modes of self-limitation. Also there, self-analysis as presented above, might not be warranted. Leadership, there, can achieve similar goals without the investment, risk, and the slowing down of governance that can accompany the more collective form of therapy.

Besides the therapeutic applications of the governance perspective proposed in this book, it can be used in a variety of other ways. Observers of public policy, administration, politics, and planning might find use for these ideas in uncovering modes of self-limitation at work in governance. Many theories, often in a modernist vein, confuse normative and analytic perspectives, which can lead to confusion between theory and method (Allmendinger, 2017; Hillier & Gunder, 2003). A method of governance is presented as a theory of governance, and an understanding of governance, often based on essentializing models, is expected to be guiding for governance practice. Or, conversely, the identification of an ideal, based on principles rather than observed practices, brings one to formulate deviations of this ideal as problems, rather than as reflecting a natural diversity in governance paths (Forester & Fischer, 1993).

Psychoanalysis was and remains of great value for re-clarifying the distinction between ideal and reality, partly, and paradoxically, by showing how fantasies always structure our version of reality, how desire, affect, and ideas always co-constitute each other, in ways that are never immediately accessible for individuals or communities. This opacity is not a fringe phenomenon, or a result of an imperfect rationality that leaves those fringes unstructured. Opacity is structural, an accompaniment for selective non-thinking, non-observation, and, crucially, non-acting and not-organizing.

Certainly, much more can be said about these opacities, which are essential for the understanding of problematic modes of self-limitation. The contribution of psychoanalysis can be explored much further. Collective affect, its association with conscious and unconscious narratives and signifiers, deserves further treatment, especially, we believe, anxiety and fear, which are behind many of the regressive tendencies in our twenty-first-century democracies (Hirvonen, 2017; Hulme, 2018). More positively, the abundance of modes and mechanisms of sublimation at work in governance and the conditions for their encouragement, warrant a more detailed investigation (Kim et al., 2013). The productive and less productive tensions between governance and community, and the need to forget the act of self-creation with each decision are all worthy of much more extensive study (Picione & Lozzi, 2021).

The vagaries of both psychoanalysis and political and governance theory over the last century, the decade-long backlash against psychoanalysis, and the remarkable resilience of modernist perspectives on reality, governance, and identity have made it difficult to initiate a rapprochement, to patiently explore what psychoanalysis could contribute to the understanding of governance. Trailblazers such as Yannis Stavrakakis, Jason Glynos, Jean Hillier, Michael Gunder, and, of course, Slavoj Žižek broke ground, but often found much resistance themselves. Now, with a history of backlash against modernist versions of statecraft behind us, with at least an understanding of alternative theoretical options, an appreciation of self-organization, local knowledge and stories, participation, the growth of informal institutions, and with at least a cultural environment more receptive to psychoanalysis, the time might be ripe for a reintroduction of psychoanalysis to the interpretation of the way we organize and understand ourselves, and the way these two domains of human life interact.

Communities can remember and forget, navigate the future, orient themselves toward changing environments, delude themselves, and fantasize productively. These processes are never technocratic. They are always more than a summation of individual and group desires through formal political processes. Governance obliges communities to reinterpret themselves over and over again. Unifying discourse as well as forms of organization maintaining that unity can, but does not have to crystallize. They require maintenance. Each form of thinking and organizing that emerges, comes with unique patterns of blind spots, rigidity and flexibility, transparency and opacity. This, we firmly believe, is not a tragic assessment of the human condition, but one that gives more space to creativity, diversity, ambition, and, indeed, freedom.

References

Abse, W., & Jessner, L. (2019). The Psychodynamic Aspects of Leadership. In S. R. Graubard & G. Holton (Eds.), *The Psychodynamic Aspects of Leadership* (pp. 76–93). Columbia University Press. https://doi.org/10.7312/grau91070-006

Acemoglu, D., & Robinson, J. A. (2013). *Why Nations Fail: The Origins of Power, Prosperity, and Poverty*. Crown Business.

Allmendinger, P. (2017). *Planning Theory* (3rd ed., p. 346). Macmillan Education UK. https://books.google.ca/books?id=XQAoDwAAQBAJ

Alvesson, M., & Spicer, A. (2010). *Metaphors We Lead By: Understanding Leadership in the Real World*. Routledge.

Anderson, B., Grove, K., Rickards, L., & Kearnes, M. (2020). Slow emergencies: Temporality and the racialized biopolitics of emergency governance. *Progress in Human Geography*, *44*(4), 621–639. https://doi.org/10.1177/030913251 9849263

Antze, P., & Lambek, M. (Eds.). (2016). *Tense Past* (0 ed.). Routledge. https://doi.org/ 10.4324/9781315022222

Beunen, R., Duineveld, M., & Van Assche, K. A. M. (2016). Evolutionary Governance Theory and the Adaptive Capacity of the Dutch Planning System. In G. de Roo et al.

(Eds.), *Spatial Planning in a Complex Unpredictable World of Change* (pp. 98–116). InPlanning.

Beunen, R., & Patterson, J. J. (2017). Analysing institutional change in environmental governance: Exploring the concept of 'institutional work.' *Journal of Environmental Planning and Management, 62*(1), 12–29.

Bistoen, G., Vanheule, S., & Craps, S. (2014). Nachträglichkeit: A Freudian perspective on delayed traumatic reactions. *Theory & Psychology, 24*(5), 668–687. https://doi.org/10.1177/0959354314530812

Boin, A., Kuipers, S., & Overdijk, W. (2013). Leadership in Times of Crisis: A Framework for Assessment. In P. `t Hart & K. Tindall (Eds.), *Leadership in Times of Crisis: A Framework for Assessment.* (Vol. 18, pp. 79–91). ANU Press. http://doi.org/10.22459/FGED.09.2009

Bolden, R., Gosling, J., & Hawkins, B. (2023). *Exploring Leadership: Individual, Organizational, and Societal Perspectives.* Oxford University Press.

Brans, M., & Rossbach, S. (1997). The autopoiesis of administrative systems: Niklas Luhmann on public administration and public policy. *Public Administration, 75*(3), 417–439.

Burt, R. S., Reagans, R. E., & Volvovsky, H. C. (2021). Network brokerage and the perception of leadership. *Social Networks, 65*, 33–50. https://doi.org/10.1016/j.socnet.2020.09.002

Bush, T. (2023). Distributed leadership and micropolitics. *Educational Management Administration & Leadership, 51*(3), 529–532. https://doi.org/10.1177/17411432231156397

Casey, K., Lichrou, M., & O'Malley, L. (2017). Unveiling everyday reflexivity tactics in a sustainable community. *Journal of Macromarketing, 37*(3), 227–239. https://doi.org/10.1177/0276146716674051

Collinson, D. (2011). Critical Leadership Studies. In A. Bryman, K. Grint, & D. L. Collinson, (Eds.), *The Sage Handbook of Leadership* (pp. 181–194). SAGE.

Cooke, B., & Kothari, U. (2001). *Participation: The New Tyranny?* Zed books.

Czarniawska, B. (2014). *A Theory of Organizing.* Edward Elgar Publishing.

Czarniawska, B., & Mazza, C. (2003). Consulting as a Liminal space. *Human Relations, 56*(3), 267–290. https://doi.org/10.1177/0018726703056003612

Demertzis, N. (2019). Populisms and Emotions 1. In P. Cossarini, & F. Vallespín (Eds.), *Populism and Passions* (pp 31–48). Routledge.

Elwood, S. (2004). Partnerships and participation: Reconfiguring urban governance in different state contexts. *Urban Geography, 25*(8), 755–770. https://doi.org/10.2747/0272-3638.25.8.755

Eyers, T. (2012). *Lacan and the Concept of the "Real."* Palgrave Macmillan.

Fink, B. (1999). *A Clinical Introduction to Lacanian Psychoanalysis: Theory and Technique.* Harvard University Press.

Fink, B. (2013a). *Against Understanding, Volume 1: Commentary and Critique in a Lacanian Key.* Routledge.

Fink, B. (2013b). *Against Understanding, Volume 2: Cases and Commentary in a Lacanian Key.* Routledge. https://doi.org/10.4324/9781315884035

Fink, B. (2017a). *A Clinical Introduction to Freud: Techniques for Everyday Practice.* W. W. Norton & Company.

Fink, B. (2017b). *Lacan on Love: An Exploration of Lacan's Seminar VIII, Transference.* John Wiley & Sons.

Forester, J., & Fischer, F. (1993). *The Argumentative Turn in Policy Analysis and Planning.* Duke University Press.

Freud, S. (2003). *The Uncanny* (H. Haughton, Ed.; D. McLintock, Trans.; Illustrated edition). Penguin Classics.

Freud, S. (2015). *Civilization and Its Discontents.* Broadview Press.

Giesen, B., & Eisenstadt, S. N. (2015). *Triumph and Trauma.* Routledge. https://doi.org/ 10.4324/9781315631455

Glynos, J. (2011). On the ideological and political significance of fantasy in the organization of work. *Psychoanalysis, Culture & Society, 16*(4), 373–393. https://doi.org/ 10.1057/pcs.2010.34

Glynos, J., & Stavrakakis, Y. (2008). Lacan and political subjectivity: Fantasy and enjoyment in psychoanalysis and political theory. *Subjectivity, 24*(1), 256–274. https://doi.org/10.1057/sub.2008.23

Grint, K., & Jones, O. S. (2022). *Leadership: Limits and Possibilities.* Bloomsbury Publishing.

Gunder, M., & Hillier, J. (2009). *Planning in Ten Words or Less: A Lacanian Entanglement with Spatial Planning.* Ashgate Publishing, Ltd.

Gunn, S., & Hillier, J. (2014). When uncertainty is interpreted as risk: An analysis of tensions relating to spatial planning reform in England. *Planning Practice & Research, 29*(1), 56–74. https://doi.org/10.1080/02697459.2013.848530

Helsloot, I., & Groenendaal, J. (2017). It's meaning making, stupid! Success of public leadership during flash crises. *Journal of Contingencies and Crisis Management, 25*(4), 350–353. https://doi.org/10.1111/1468-5973.12166

Hillier, J. (2003). Puppets of populism? *International Planning Studies, 8*(2), 157–157.

Hillier, J., & Gunder, M. (2003). Planning fantasies? An exploration of a potential Lacanian framework for understanding development assessment planning. *Planning Theory, 2*(3), 225–248. https://doi.org/10.1177/147309520323005

Hirvonen, A. (2017). Fear and anxiety: The nationalist and racist politics of fantasy. *Law and Critique, 28*(3), 249–265. https://doi.org/10.1007/s10978-017-9210-y

Hoggett, P. (2006). Conflict, ambivalence, and the contested purpose of public organizations. *Human Relations, 59*(2), 175–194. https://doi.org/10.1177/00187 26706062731

Hoggett, P. (2015). *Politics, Identity and Emotion.* Routledge. https://doi.org/10.4324/ 9781315632704

Holstein, J. A., & Gubrium, J. F. (1999). *The Self We Live By: Narrative Identity in a Postmodern World.* Oxford University Press.

Hook, D. (2011). *A Critical Psychology of the Postcolonial: The Mind of Apartheid.* Routledge. https://doi.org/10.4324/9780203140529

Hulme, M. (2009). *Why We Disagree about Climate Change: Understanding Controversy, Inaction and Opportunity.* Cambridge University Press.

Hulme, M. (2018). "Gaps" in climate change knowledge: Do they exist? Can they be filled? *Environmental Humanities, 10*(1), 330–337. https://doi.org/10.1215/22011 919-4385599

Jarrett, M., & Vince, R. (2023). Mitigating anxiety: The role of strategic leadership groups during radical organisational change. *Human Relations, 77*(8), 1178–1208. 00187267231169143. https://doi.org/10.1177/00187267231169143

Jarzabkowski, P., & Seidl, D. (2008). The Role of meetings in the social practice of strategy. *Organization Studies, 29*(11), 1391–1426. https://doi.org/10.1177/01708 40608096388

John, B., Keeler, L. W., Wiek, A., & Lang, D. J. (2015). How much sustainability substance is in urban visions? – An analysis of visioning projects in urban planning. *Cities, 48*, 86–98. https://doi.org/10.1016/j.cities.2015.06.001

Jun, J. S. (2012). *The Social Construction of Public Administration: Interpretive and Critical Perspectives*. State University of New York Press.

Kapoor, I., Fridell, G., Sioh, M., & Vries, P. de. (2023). *Global Libidinal Economy*. State University of New York Press.

Kim, E., Zeppenfeld, V., & Cohen, D. (2013). Sublimation, culture, and creativity. *Journal of Personality and Social Psychology, 105*(4), 639–666. https://doi.org/10.1037/a0033487

Luhmann, N. (1987). *Soziale Systeme: Grundriss einer allgemeinen Theorie*. Suhrkamp.

Luhmann, N. (1990). *Political Theory in the Welfare State*. De Gruyter.

Luhmann, N. (1995). *Social Systems* (Vol. 1). Stanford University Press Stanford.

Luhmann, N. (1997). The control of intransparency. *Systems Research and Behavioral Science, 14*(6), 359–371. https://doi.org/10.1002/(SICI)1099-1743(199711/12)14:6<359::AID-SRES160>3.0.CO;2-R

Luhmann, N. (2018). *Organization and Decision*. Cambridge University Press.

Luhmann, N., Ziegert, K. A., & Kastner, F. (2004). *Law as a Social System*. Oxford University Press.

Magris, C. (2016). *Microcosms*. www.vlebooks.com/vleweb/product/openreader?id=none&isbn=9781446433768

Marks, M., Shearing, C., & Wood, J. (2009). Who should the police be? Finding a new narrative for community policing in South Africa. *Police Practice and Research, 10*(2), 145–155. https://doi.org/10.1080/15614260802264560

Matthies-Boon, V. (2017). Shattered worlds: Political trauma amongst young activists in post-revolutionary Egypt. *The Journal of North African Studies, 22*(4), 620–644. https://doi.org/10.1080/13629387.2017.1295855

McCann, E. J. (2001). Collaborative visioning or urban planning as therapy? The politics of public-private policy making. *The Professional Geographer, 53*(2), 207–218.

McGowan, T. (2012). *The End of Dissatisfaction?: Jacques Lacan and the Emerging Society of Enjoyment*. State University of New York Press.

McSwite, O. C. (1997). Jacques Lacan and the theory of the human subject: How psychoanalysis can help public administration. *American Behavioral Scientist, 41*(1), 43–63. https://doi.org/10.1177/0002764297041001005

Miller, H. T. (2002). *Postmodern Public Policy*. SUNY Press.

Miller, H. T. (2012). *Governing Narratives: Symbolic Politics and Policy Change*. University of Alabama Press.

Minkler, M. (2012). Introduction to Community Organizing and Community Building. In M. Minkler & N. Wallerstein (Eds.), *1. Introduction to Community Organizing and Community Building* (pp. 5–26). Rutgers University Press. https://doi.org/10.36019/9780813553146-003

Misztal, B. (2003). *Theories of Social Remembering*. McGraw-Hill Education (UK).

Mohatt, N. V., Thompson, A. B., Thai, N. D., & Tebes, J. K. (2014). Historical trauma as public narrative: A conceptual review of how history impacts present-day health. *Social Science & Medicine, 106*, 128–136. https://doi.org/10.1016/j.socscimed.2014.01.043

Moxnes, P. (1998). Fantasies and fairy tales in groups and organizations: Bion's basic assumptions and the deep roles. *European Journal of Work and Organizational Psychology, 7*(3), 283–298. https://doi.org/10.1080/135943298398718

North, D. C., Wallis, J. J., & Weingast, B. R. (2009). *Violence and Social Orders: A Conceptual Framework for Interpreting Recorded Human History.* Cambridge University Press.

Picione, R. D. L., & Lozzi, U. (2021). Uncertainty as a constitutive condition of human experience: Paradoxes and complexity of sensemaking in the face of the crisis and uncertainty. *International Journal of Psychoanalysis and Education: Subject, Action & Society, 1*(2), Article 2. https://doi.org/10.32111/SAS.2021.1.2.2

Pottage, A., & Sherman, B. (2010). *Figures of Invention: A History of Modern Patent Law.* Oxford University Press. https://doi.org/10.1093/acprof:osobl/9780199595631.001.0001

Rasche, A., & Seidl, D. (2020). A Luhmannian perspective on strategy: Strategy as paradox and meta-communication. *Critical Perspectives on Accounting, 73,* 101984–101984. https://doi.org/10.1016/j.cpa.2017.03.004

Renström, E. A., Bäck, H., & Carroll, R. (2023). Threats, emotions, and affective polarization. *Political Psychology, 44*(6), 1337–1366. https://doi.org/10.1111/pops.12899

Rutan, S. J. (1999). Psychoanalytic Group Psychotherapy*. In J. R. Price, D. R. Hescheles, A. Rae Price, & A. R. Price (Eds.), *A Guide to Starting Psychotherapy Groups* (pp. 151–166). Academic Press. https://doi.org/10.1016/B978-012564745-8/50019-4

Sabbadini, A. (1992). Listening to silence. *The Scandinavian Psychoanalytic Review. 15,* 27–36. www.tandfonline.com/doi/abs/10.1080/01062301.1992.10592268

Sandler, J. (with International Psycho-Analytical Association). (1991). *On Freud's "Analysis Terminable and Interminable."* Yale University Press.

Scheidlinger, S. (1968). The concept of regression – Group Psychotherapy. *International Journal of Group Psychotherapy, 18,* 30–20. www.tandfonline.com/doi/abs/10.1080/00207284.1968.11508332

Scott, J. C. (1998). *Seeing Like A State: How Certain Schemes to Improve the Human Condition Have Failed.* Yale University Press.

Seidl, D. (2016). *Organisational Identity and Self-Transformation: An Autopoietic Perspective.* Routledge.

Shields, R. (2005). *The Virtual.* Routledge.

Shipley, R., & Utz, S. (2012). Making it count: A review of the value and techniques for public consultation. *Journal of Planning Literature, 27*(1), 22–42. https://doi.org/10.1177/0885412211413133

Stavrakakis, Y. (2002). *Lacan and the Political.* Routledge.

Stavrakakis, Y. (2010). Symbolic Authority, Fantasmatic Enjoyment and the Spirits of Capitalism: Genealogies of Mutual Engagement. In C. Cederström & C. Hoedemaekers (Eds.), *Lacan and Organization* (pp. 59–100).

Swyngedouw, E. (2022). The unbearable lightness of climate populism. *Environmental Politics, 31*(5), 904–925. https://doi.org/10.1080/09644016.2022.2090636

Tourish, D. (2013). *The Dark Side of Transformational Leadership: A Critical Perspective.* Routledge.

Vaara, E., Sonenshein, S., & Boje, D. (2016). Narratives as sources of stability and change in organizations: Approaches and directions for future research. *Academy of Management Annals, 10*(1), 495–560. https://doi.org/10.5465/19416520.2016.1120963

Valentinov, V., Verschraegen, G., & Van Assche, K. (2019). The limits of transparency: A systems theory view. *Systems Research and Behavioral Science, 36*(3), 289–300. https://doi.org/10.1002/sres.2591

Van Assche, K., Beunen, R., & Gruezmacher, M. (2024). *Strategy for Sustainability Transitions: Governance, Community and Environment*. Edward Elgar Publishing.

Van Assche, K., Beunen, R., Gruezmacher, M., Duineveld, M., Deacon, L., Summers, R., Hallstrom, L., & Jones, K. (2019). Research methods as bridging devices: Path and context mapping in governance. *Journal of Organizational Change Management, ahead-of-print* (ahead-of-print). https://doi.org/10.1108/JOCM-06-2019-0185

Van Assche, K., Greenwood, R., & Gruezmacher, M. (2022). The local paradox in grand policy schemes. Lessons from Newfoundland and Labrador. *Scandinavian Journal of Management, 38*(3), 101212. https://doi.org/10.1016/j.scaman.2022.101212

Van Assche, K., Gruezmacher, M., & Deacon, L. (2020). Land use tools for tempering boom and bust: Strategy and capacity building in governance. *Land Use Policy, 93*, 103994–103994. https://doi.org/10.1016/j.landusepol.2019.05.013

Van Assche, K., Gruezmacher, M., Marais, L., & Perez-Sindin, X. (2023). *Resource Communities: Past Legacies and Future Pathways*. Taylor & Francis. https://books.google.ca/books?id=4EHUEAAAQBAJ

Vanheule, S. (2016). Capitalist discourse, subjectivity and Lacanian psychoanalysis. *Frontiers in Psychology, 7*, 1948. https://doi.org/10.3389/fpsyg.2016.01948

Vélez-Torres, I., Gough, K., Larrea-Mejía, J., Piccolino, G., & Ruette-Orihuela, K. (2022). "Fests of vests": The politics of participation in neoliberal peacebuilding in Colombia. *Antipode, 54*(2), 586–607. https://doi.org/10.1111/anti.12785

Walker, G., & Shove, E. (2007). Ambivalence, sustainability and the governance of socio-technical transitions. *Journal of Environmental Policy & Planning, 9*(3–4), 213–225. https://doi.org/10.1080/15239080701622840

Wolf, A., & Schwartz, E. K. (1962). *Psychoanalysis in Groups*. Grune and Stratton.

Yanow, D. (2000). *Conducting Interpretive Policy Analysis* (Vol. 47). Sage.

Žižek, S. (2009). *In Defense of Lost Causes*. Verso. http://public.ebookcentral.proquest.com/choice/publicfullrecord.aspx?p=5176960

Žižek, S. (2019). *The Sublime Object of Ideology*. Verso Books.

Žižek, S. (2020). *The Plague of Fantasies*. Verso Books.

Index

For Product Safety Concerns and Information please contact our EU
representative GPSR@taylorandfrancis.com
Taylor & Francis Verlag GmbH, Kaufingerstraße 24, 80331 München, Germany

www.ingramcontent.com/pod-product-compliance
Lightning Source LLC
Chambersburg PA
CBHW070329270326
41926CB00017B/3821

9 781032 696683